大学应用型课程专业（精品）系列教材　　喻世友◎主编

大学应用型课程专业（精品）系列教材·计算机应用类　　杨智◎主编

Java 程序设计实训
——增量式项目驱动一体化教程

苑俊英　杨　智　主编 / 谭志国　陈海山　温泉思　副主编

中山大学出版社
SUN YAT-SEN UNIVERSITY PRESS
·广州·

版权所有　翻印必究

图书在版编目（CIP）数据

Java 程序设计实训：增量式项目驱动一体化教程/苑俊英，杨智主编；谭志国，陈海山，温泉思副主编. —广州：中山大学出版社，2015.8

［大学应用型课程专业（精品）系列教材/喻世友主编；大学应用型课程专业（精品）系列教材·计算机应用类/杨智主编］

ISBN 978 - 7 - 306 - 05373 - 2

Ⅰ.①J… Ⅱ.①苑… ②杨… Ⅲ.①JAVA 语言—程序设计—教材 Ⅳ.①TP312

中国版本图书馆 CIP 数据核字（2015）第 170315 号

出版人：	徐　劲
责任编辑：	黄浩佳
封面设计：	曾　斌
责任校对：	王　睿
责任技编：	何雅涛

出版发行：中山大学出版社
电　　话：编辑部 020 - 84111996，84113349，84111997，84110779
　　　　　发行部 020 - 84111998，84111981，84111160
地　　址：广州市新港西路 135 号
邮　　编：510275　　　传　真：020 - 84036565
网　　址：http://www.zsup.com.cn　　E-mail：zdcbs@mail.sysu.edu.cn
印　刷　者：虎彩印艺股份有限公司
规　　格：787mm×1092mm　1/16　22.75 印张　526 千字
版次印次：2015 年 8 月第 1 版　2015 年 8 月第 1 次印刷
定　　价：46.00 元

如发现本书因印装质量影响阅读，请与出版社发行部联系调换

大学应用型课程专业（精品）系列教材
编 委 会

主 任：喻世友

委 员：（按姓氏拼音排序）

陈功玉　陈剑波　陈天祥　丁建新　方海云　冯　原
何江海　黄静波　黎颂文　廖俊平　孙　立　王丽荣
卫建国　杨　智　喻世友　赵过渡

大学应用型课程专业（精品）系列教材·计算机应用类
编 委 会

主　编　杨　智

副主编　苑俊英

编　委　（按姓氏拼音排序）

薄　宏　陈海山　曹惠茹　成海秀　戴宏明　洪维恩
卢洪斌　谭志国　温泉思　杨　智　苑俊英　张鉴新
钟晓婷

本书编委会

主　编　苑俊英　杨　智

副主编　谭志国　陈海山　温泉思

编　委　（按姓氏拼音排序）

陈海山　谭志国　温泉思　杨　智　苑俊英

前　　言

程序设计课程在计算机相关专业的教学体系中占据非常重要的作用,学好程序设计对软件工程、数据库、高级编程以及综合实训等课程有很大帮助,学习程序设计课程要注重程序设计方法与程序设计语言两方面内容的学习。随着互联网的发展和广泛应用,Java 语言伴随着分布式程序设计方法和技术的需求问世,Java 语言具有面向对象、与平台无关、多线程、稳定性和安全性等特性,为用户提供了一个良好的程序设计环境,尤其适合互联网的应用开发,拥有全球最大的开发者专业社群。Java 语言是当今最具代表性的面向对象编程语言之一,也是实际软件项目开发中使用的主流语言之一,在全球云计算和移动互联网产业蓬勃发展的环境下,具有显著优势和广阔前景。

在应用型人才培养模式下,Java 程序设计及实训课程的教学过程应该注重学生实践能力的培养。由于学时的限制,教师在课堂如何组织教学,如何安排教学内容,采用什么样的教学方法,都对教学效果起着举足轻重的作用。所以,编者根据自身的实践及教学经验,在 Java 程序设计实训课程教学中提出了一体化增量式项目驱动的教学方法。在教学内容安排上分两个阶段:专项实训和综合实训。在专项实训部分以一个小型项目贯穿整个教学过程,教师采取增量式的项目开发方法安排教学和实训内容,通过专项实训的增量迭代让学生体会项目的开发流程、开发方法和技术的使用,最终掌握如何在一体化的教学模式下,综合运用各专项实训技能和知识完成一个小型项目的设计和实现过程。在综合实训部分安排几个中型项目,学生以团队的形式选取一个综合项目进行实践,在该阶段重点培养学生的团队协作能力、沟通能力、分析问题和解决问题的能力、技术和知识的综合运用能力以及文档的撰写能力。

本书为广东省教育厅"育苗工程(自然科学)"之"计算思维与应用型本科人才培养结合下的计算机专业基础课程建设"项目成果之一,是《Java 程序设计及应用——增量式项目驱动一体化教程》(ISBN 978-7-121-20764-8)的后续实训课程的配套教材,全书共分两部分,共 9 章,每章包含一个参考项目增量任务的实现过程以及一个实训项目的任务安排,从需求分析、参考案例实现过程到任务安排与实现,深入浅出地进行介绍。本书注重增量式的软件设计过程和一体化的项目驱动案例分析与实现,读者学习每章后,可立即动手实践,在实践过程中巩固知识和技能,并学习增量式项目开发。

一体化的增量式项目驱动是本书的特色,本书在实训项目中,给出了参考项目的框架性代码,读者通过分析与理解,能融会贯通,进行其他项目程序的编写工作。本书全部案例都是编者多年开发经验和教学经验的积累和总结,具有代表性。

本书由苑俊英、杨智、谭志国、陈海山、温泉思共同编写,主要编写人员和分工如下:第 1 章至第 4 章、第 8 章由苑俊英、杨智编写;第 5 章、第 6 章由温泉思编写;第

7 章由陈海山编写；第 9 章由谭志国编写。全书由杨智、苑俊英进行统稿和定稿。本书中综合实训部分的代码由中山大学南方学院学生蔡力能、朱彦瑾进行了测试工作。

在本书编写过程中，河北大学袁方教授、中山大学南方学院洪维恩教授、中山大学南方学院杨秀珍副教授提出了许多宝贵的建议和意见，在此表示衷心感谢。

由于作者水平有限，书中难免存在疏漏，恳请读者批评指正。本书还配有教学课件和实训案例的参考代码，有需要的读者可与作者联系，邮箱为 cihisa@126.com。

编者
2015 年 3 月

目　　录

第一部分　专项实训 ……………………………………………………………………… 1

第 1 章　开发环境 ………………………………………………………………………… 3
1.1　Java 语言介绍 ……………………………………………………………………… 3
1.2　Java 开发工具包 JDK ……………………………………………………………… 3
1.3　集成开发环境 Eclipse ……………………………………………………………… 9
1.4　集成开发环境 MyEclipse ………………………………………………………… 10
1.5　数据库软件 MySQL ……………………………………………………………… 11
实验 01　开发环境搭建与使用 ……………………………………………………… 22

第 2 章　专项实训需求分析 …………………………………………………………… 23
2.1　车票预订系统用户及界面分析 …………………………………………………… 23
　　2.1.1　车票预订系统用例分析 ………………………………………………… 24
　　2.1.2　车票预订系统数据分析 ………………………………………………… 25
2.2　专项实训增量划分及进度安排 …………………………………………………… 26
实验 02　专项实训需求分析 ………………………………………………………… 26

第 3 章　GUI 编程及事件处理 ………………………………………………………… 28
3.1　车票预定系统的界面设计与实现 ………………………………………………… 28
　　3.1.1　用户登录界面 …………………………………………………………… 28
　　3.1.2　班车查询及预订界面 …………………………………………………… 31
　　3.1.3　用户确认订票信息界面 ………………………………………………… 35
3.2　事件处理（以 ActionListener 为例） …………………………………………… 38
　　3.2.1　用户登录事件处理 ……………………………………………………… 38
　　3.2.2　班车查询及订票事件处理 ……………………………………………… 41
　　3.2.3　确认订票信息事件处理 ………………………………………………… 46
实验 03　GUI 编程 …………………………………………………………………… 48
实验 04　事件处理 …………………………………………………………………… 49

第 4 章　数据库编程 …………………………………………………………………… 50
4.1　数据库设计 ………………………………………………………………………… 50
　　4.1.1　E－R 图 ………………………………………………………………… 50
　　4.1.2　数据库表设计 …………………………………………………………… 50
　　4.1.3　数据库脚本 ……………………………………………………………… 52
4.2　JDBC 配置 ………………………………………………………………………… 53
4.3　登录功能的数据库实现 …………………………………………………………… 54

  4.3.1 用户登录时序图 ·································· 54
  4.3.2 功能实现 ·· 55
 4.4 班车查询功能的数据库实现 ································ 56
  4.4.1 班车查询功能时序图 ·································· 56
  4.4.2 功能实现 ·· 56
 4.5 班车预定功能的数据库实现 ································ 59
  4.5.1 班车预定功能时序图 ·································· 59
  4.5.2 功能实现 ·· 60
 实验 05 数据库编程 ·· 61

第 5 章 网络通信编程 ·· 62
 5.1 网络通信技术分析 ·· 62
  5.1.1 网络编程的基本知识 ·································· 62
  5.1.2 URL 通信方式 ······································ 63
  5.1.3 套接字通信方式 ···································· 64
 5.2 TCP 对象数据流通信功能的设计与搭建 ···················· 66
  5.2.1 对象序列化的实现 ·································· 66
  5.2.2 对象数据流通信功能的搭建 ·························· 68
 5.3 TCP 对象数据流通信功能的实现 ···························· 77
  5.3.1 TCP 用户登录功能实现 ······························ 77
  5.3.2 TCP 班车查询及订票功能实现 ························ 80
 实验 06 网络通信编程 ·· 83

第 6 章 多线程编程 ·· 85
 6.1 车票预定系统的多线程创建 ································ 85
 6.2 车票预定系统的线程同步问题 ······························ 91
 实验 07 多线程编程 ·· 93

第二部分 综合实训 ·· 95
第 7 章 办公管理系统 ·· 97
 7.1 需求分析与项目目标 ······································ 97
  7.1.1 用例分析 ·· 98
  7.1.2 需求分析 ·· 100
  7.1.3 项目目标 ·· 101
 7.2 功能分析与软件设计 ······································ 101
  7.2.1 界面效果设计 ······································ 102
  7.2.2 登录功能 ·· 102
  7.2.3 管理模块 ·· 103
  7.2.4 考勤模块 ·· 108
  7.2.5 公告模块 ·· 112

		7.2.6	通信模块	113
		7.2.7	数据库表格与脚本设计	116
		7.2.8	增量开发计划	120
	7.3	项目增量开发		121
		7.3.1	增量 7-1：搭建系统主体架构	121
		7.3.2	增量 7-2：信息管理模块	140
		7.3.3	增量 7-3：实现公告	168
		7.3.4	增量 7-4：实现考勤	184
	7.4	总结与回顾		196
	7.5	软件使用说明		197
		7.5.1	开发环境	197
		7.5.2	软件使用方法	198
	7.6	本章小结		198

第8章 Web 考勤系统 · 199

	8.1	需求分析与项目目标		199
		8.1.1	用例分析	199
		8.1.2	业务流分析	201
	8.2	界面设计		203
		8.2.1	用户登录界面	203
		8.2.2	学生界面	205
		8.2.3	教师界面	205
		8.2.4	管理员界面	206
		8.2.5	界面效果	206
		8.2.6	界面设计技术——响应式设计	219
	8.3	软件设计		222
		8.3.1	系统架构设计	222
		8.3.2	基础设施层	223
		8.3.3	领域层	229
		8.3.4	应用层	237
		8.3.5	用户接口层	239
		8.3.6	数据库设计	246
	8.4	系统实现		254
		8.4.1	用户界面关键算法实现	254
		8.4.2	领域层实现	256
		8.4.3	用户接口层实现	273
		8.4.4	应用层实现	290
	8.5	系统使用说明		291
		8.5.1	运行环境	291

　　　　8.5.2　单应用服务器部署 …………………………………………………… 291
　　8.6　本章小结 ……………………………………………………………………… 292
第9章　基于Android平台的视频播放器 …………………………………………… 293
　　9.1　Android简介 ………………………………………………………………… 293
　　　　9.1.1　Android系统架构 ……………………………………………………… 294
　　　　9.1.2　Android应用程序组成 ………………………………………………… 297
　　　　9.1.3　搭建Android开发环境 ………………………………………………… 301
　　　　9.1.4　开发第一个Android应用程序 ………………………………………… 302
　　9.2　视频播放器开发 ……………………………………………………………… 305
　　　　9.2.1　需求分析 ………………………………………………………………… 307
　　　　9.2.2　项目目标 ………………………………………………………………… 310
　　　　9.2.3　增量开发计划 …………………………………………………………… 310
　　9.3　项目增量开发 ………………………………………………………………… 311
　　　　9.3.1　增量9-1：搜索视频并列表显示 ……………………………………… 311
　　　　9.3.2　增量9-2：视频播放及控制功能的实现 ……………………………… 333
　　　　9.3.3　增量9-3　自定义视频播放界面 ……………………………………… 336
　　9.4　总结与回顾 …………………………………………………………………… 349
　　9.5　软件使用说明 ………………………………………………………………… 350
　　　　9.5.1　开发环境 ………………………………………………………………… 350
　　　　9.5.2　软件使用方法 …………………………………………………………… 350
　　9.6　本章小结 ……………………………………………………………………… 350

参考文献 …………………………………………………………………………………… 351

第一部分　专项实训

　　本教材第一部分主要按照增量式的项目开发流程设计并实现一个小型项目，教材的参考案例模版为车票预订系统。要求学生重点掌握编程环境搭建、GUI 编程、事件处理、数据库编程、网络编程和多线程等知识和技术的熟练应用，并能融会贯通，体会参考案例增量的设计步骤和实现框架，在此基础上能独立开发类似的 Java 项目。

第1章 开发环境

1.1 Java 语言介绍

Java 语言是由 Sun 公司开发而出的新一代编程语言（已被 Oracle 收购），诞生于 1995 年，是在 C 语言和 C++ 语言基础上创建的，最初用于开发电冰箱、电烤箱等消费类电子产品市场，目前已广泛应用于开发各种网络应用软件，成为最流行的程序设计语言之一。

Java 平台由 Java 虚拟机（Java Virtual Machine，简称 JVM）和 Java 应用程序编程接口（Application Programming Interface，简称 API）构成，J2SE 平台（Java 的标准版）的版本也不断地发展，从 1996 年的 JDK 1.0 发展到目前的最新版本 JDK 8.0。Sun 公司免费提供的开发工具包的早期版本简称为 JDK（Java Developer's Kit），现在推出的开发工具包分化成为 3 个版本：J2SE（Java2 Platform，Standard Edition）、J2EE（Java2 Platform，Enterprise Edition）和 J2ME（Java2 Platform，Micro Edition）。J2SE 是用于工作站和个人计算机的标准开发工具包，J2EE 是用于企业级开发的工具包，J2ME 是用于开发智能消费型电子产品。

Java 作为一种高级程序设计语言，与其他高级语言相比，最重要的特点是它与平台的无关性，也就是常说的"Write once run anywhere"。另外，Java 具有简单、面向对象、与平台无关性和可移植性、可靠性和安全性、多线程并发机制、分布式、动态的内存管理机制等特点。

1.2 Java 开发工具包 JDK

Java 开发工具包 JDK 是一个编写 Java 程序的开发环境，JDK 工具包可在 http://java.sun.com 免费下载，目前 JDK 版本 8.0。下载时要注意自己计算机的操作系统类型，下载的安装程序应当与自己计算机的操作系统相匹配，本教材使用的 JDK 版本为 jdk-8u20-windows-i586。

下载完成后，即可进行安装，安装过程只要遵循安装程序的指示进行即可。具体安装步骤如下所示：

（1）双击 JDK 安装文件，即可进入 JDK 安装向导界面，如图 1-1 所示。

3

图1-1　安装向导界面

（2）单击【下一步】进入自定义安装界面，在该界面可以修改安装路径，如图1-2所示。

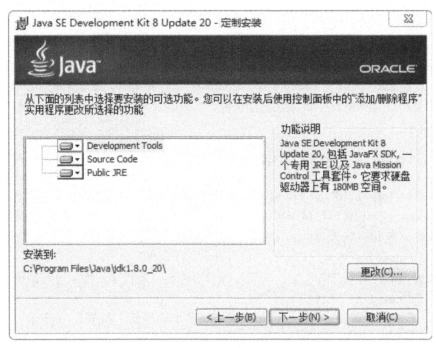

图1-2　自定义安装界面

(3) 进行设置后,单击【下一步】即可安装,安装过程如图 1-3 所示。

图 1-3　安装进度

(4) 修改 JRE 安装路径,如图 1-4 所示,可单击【更改】按钮进行路径的更改。

图 1-4　安装 JRE

(5) 修改完成后，单击【下一步】开始安装 JRE。图 1-5 为 JPE 安装过程，当进入图 1-6 所示界面时安装完成。

图 1-5　正在安装 JRE

图 1-6　安装完成

成功安装 JDK 后，需要手动配置 JDK 环境变量。一共需要设置三个环境变量：JAVA_HOME、CLASSPATH、PATH。值得注意的是，不同操作系统下配置过程稍有不同，本教材以 Windows 7 为例进行介绍，具体配置过程如下：

（1）右击程序或桌面上的"我的电脑"，在弹出的快捷菜单中选择"属性"命令，然后在弹出的窗口左侧选择"高级系统设置"，打开【系统属性】对话框，如图 1-7 所示。

图 1-7　【系统属性】对话框

（2）在"高级"选项卡，单击【环境变量】按钮，打开【环境变量】对话框，如图 1-8 所示。

图 1-8　【环境变量】对话框

（3）单击【新建】按钮，弹出【新建系统变量】对话框，在对话框中填入如图1-9所示新建变量JAVA_HOME及变量值，其中变量值为JDK安装路径。单击【确定】按钮即可返回。

图1-9　JAVA_HOME

（4）再次单击【新建】按钮，新建CLASSPATH变量并设置变量值，如图1-10所示。CLASSPATH的值有两个，一个为当前路径（用"."表示），一个为J2SE类库所在路径，两个路径之间用分号间隔。

图1-10　CLASSPATH

（5）单击图1-8中的【编辑】按钮，修改PATH路径，向已有的PATH路径中添加bin文件路径，如图1-11所示。

图1-11　修改PATH

（6）设置完成后，连续单击【确定】按钮完成环境变量的设置。

（7）选择"开始→运行"命令，输入"cmd"后，进入MS-DOS窗口，在DOS窗口中输入javac、java命令，出现如图1-12和图1-13所示的javac或者java命令的使用方法，表示环境变量已经配置成功。

图 1-12　javac 命令使用方法

图 1-13　java 命令使用方法

1.3　集成开发环境 Eclipse

Eclipse 属于绿色软件，不需要运行安装程序，也不需要向 Windows 注册表填写信息，只需要将 Eclipse 压缩包解压就可以运行。

Eclipse 软件包可以到官方网站（http://www.eclipse.org/downloads/）下载，它可以安装在各种操作系统上，在 Windows 下安装 Eclipse，除了需要 Eclipse 软件包以外，还需要 Java 的 JDK 来支持 Eclipse 的运行,并需要设置相关的环境变量,关于 JDK 的安装及环境变量设置可参考 1.2 节。运行 Eclipse 后会弹出图 1-14 所示窗口,可在其中设置 Eclipse 程序的存放路径。

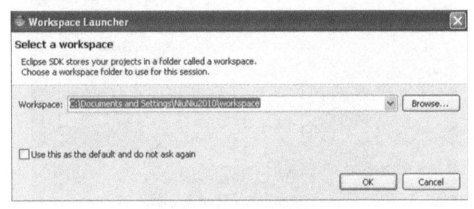

图 1-14 设置 Eclipse 存放路径

1.4 集成开发环境 MyEclipse

MyEclipse 企业级工作平台(MyEclipse Enterprise Workbench,MyEclipse)是对 EclipseIDE 的扩展,是一款商业的基于 Eclipse 的 Java EE 集成开发工具,不是免费产品,可到其官方网站(http://www.myeclipseide.com/)下载。

MyEclipse 的功能非常强大,支持也十分广泛,尤其支持各种开源产品。MyEclipse 目前支持 Java Servlet、AJAX、JSP、JSF、Struts、Spring、Hibernate、EJB3、JDBC 数据库链接工具等多项功能。可以说 MyEclipse 是几乎囊括了目前所有主流开源产品的专属 Eclipse 开发工具。

下载 MyEclipse 后即可进行安装,安装过程可自行完成。安装完成后,启动 MyEclipse,会弹出如图 1-15 所示对话框,可以设置 MyEclipse 程序的存放路径。

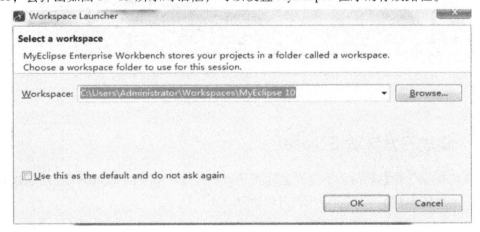

图 1-15 设置 MyEclipse 存放路径

1.5 数据库软件 MySQL

目前，MySQL 已经是各大互联网网站的首选数据库，MySQL 不仅是开源数据库管理系统，更因为它良好的性能和插件式的存储引擎，受到越来越多人的青睐。读者可到 MySQL 的官方网站（http：//dev.mysql.com/downloads/）下载使用，本教材使用 MySQL 版本为 MySQL 5.5.21。

下载完成后可进行安装，具体的安装步骤如下：

（1）首先单击 MySQL 5.5.21 的安装文件，出现安装向导界面，如图 1-16 所示，在图中单击【Next】按钮。

图 1-16　安装向导

（2）在图 1-17 中，选中【接受安装协议】并单击【Next】按钮继续安装。

（3）在图 1-18 中，选择安装类型：【Typical】（默认）、【Complete】（完全）和【Custom】（用户自定义）三个选项，选择【Custom】后单击【Next】按钮继续。

（4）在图 1-19 的自定义安装界面中选择 MySQL 数据库的安装路径，设置完之后，单击【Next】按钮继续。

（5）在图 1-20 所示的准备安装的界面中，确认前述设置，如果有误，单击【Back】按钮返回修改，否则单击【Install】按钮继续。

图 1-17 用户安装协议

图 1-18 选择安装类型

图 1-19 安装路径

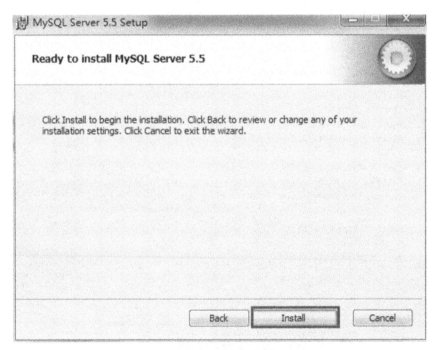

图 1-20 准备安装

(6) 出现图 1-21 和图 1-22 界面时，MySQL 安装完成。继续单击图 1-22 中的【Next】按钮对 MySQL 进行配置。

图1-21 MySQL 安装完成界面1

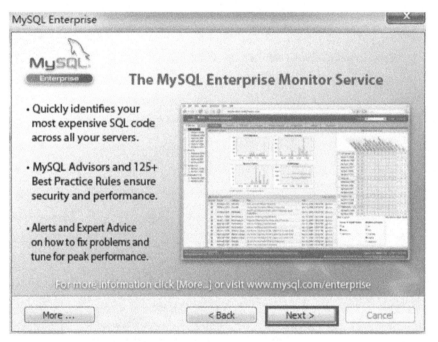

图1-22 MySQL 安装完成界面2

(7) 在图1-23所示界面中勾选【Launch the MySQL Instance Configuration Wizard】选项并单击【Finish】按钮，进入到 MySQL 配置界面，在图1-24所示界面中单击

【Next】按钮。

图 1-23　MySQL 配置 1

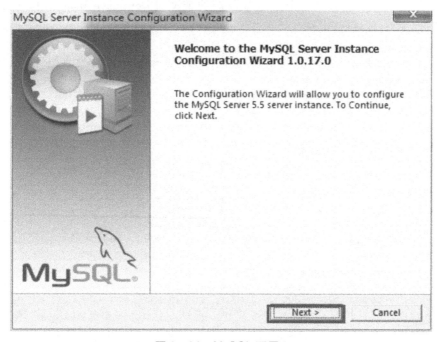

图 1-24　MySQL 配置 2

（8）在图 1-25 所示界面中选择配置的方式【Detailed Configuration（手动精确配

置)】后单击【Next】按钮继续。

图1-25 MySQL配置3

(9) 在图1-26所示界面中选择服务器的类型【Developer Machine】(开发测试类)后单击【Next】按钮继续。

图1-26 MySQL配置4

（10）在图 1-27 所示界面中选择 MySQL 数据库用途【Multifunctional Database】（通用多功能型）后单击【Next】按钮继续。

图 1-27　MySQL 配置 5

（11）在图 1-28 所示界面中对 InnoDB Tablespace 进行配置，为 InnoDB 数据库文件选择一个存储空间，要记住存储空间位置，在重装 MySQL 时要选择同一个位置，否则可能会造成数据库损坏。

图 1-28　MySQL 配置 6

17

(12) 在图 1 – 29 中所示界面设置 MySQL 的访问量及同时连接数,这里选择【Manual Setting】(Concurrent connections:15)后单击【Next】按钮继续。

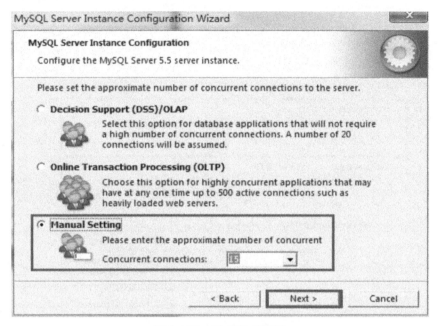

图 1 – 29　MySQL 配置 7

(13) 在图 1 – 30 所示界面中设置是否启用 TCP/IP 连接,并设置端口号,默认端口是 3306,并启用严格的语法设置,如果后面出现错误,可以将【Add firewall exception for this port】选项选中,单击【Next】按钮继续。

图 1 – 30　MySQL 配置 8

(14) 在图 1-31 所示界面中设置 MySQL 的字符编码：utf8 编码，并单击【Next】按钮继续。

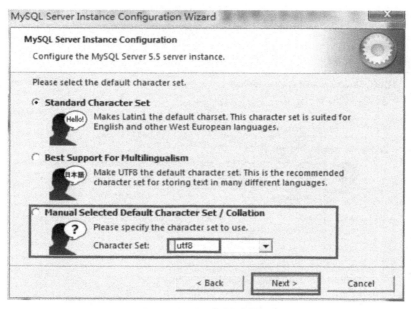

图 1-31　MySQL 配置 9

(15) 在图 1-32 所示界面中选择：是否将 MySQL 安装为 Windows 服务，还可以指定 Service Name（服务标识名称），是否将 MySQL 的 bin 目录加入到 Windows PATH（加入后，就可以直接使用 bin 下的文件，而不用指出目录名，例如，连接：mysql – u username – p password），单击【Next】按钮继续。

图 1-32　MySQL 配置 10

（16）在图1-33所示界面中设置：是否要修改默认root用户（超级管理员）的密码（默认为空），设置后单击【Next】按钮，出现图1-34所示界面时配置结束，单击【Execute】按钮执行配置，当出现1-35所示界面时配置成功并单击【Finish】按钮结束。

图1-33　MySQL配置11

图1-34　MySQL配置12

图 1-35　MySQL 配置 13

（17）在图 1-36 所示的命令窗口中输入 "mysql - u root - p" 或 "mysql - u root - p 密码" 即可启动 MySQL。

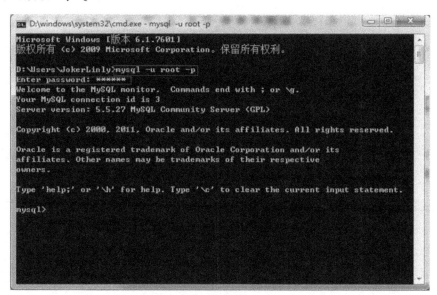

图 1-36　MySQL 启动界面

实验 01　开发环境搭建与使用

实验目的

（1）掌握 Java 开发环境的搭建过程。

（2）掌握 Java 编程的基本步骤。

（3）熟练 Eclipse、MyEclipse、MySQL 等的安装、配置及使用。

（4）掌握 Java 应用程序的创建和调试过程。

专项实训题目

（1）学生请假管理系统。

（2）其他。

实验内容

（1）JDK 的下载、安装、环境变量设置。

（2）Eclipse、MyEclipse、MySQL 等软件的安装及基本使用。

（3）Java 应用程序的创建和调试。

实验安排

开发环境搭建及编辑 Java 应用程序：3 学时。

实验报告

按照规定格式提交本次实验内容的结果。

第 2 章　专项实训需求分析

本教材引入"车票预订系统"作为专项实训的实训代码模板，在实训中应用 Java 的不同知识点及基本技能实现预期项目的不同功能需求，通过增量式的项目开发过程，最终实现一个 C/S 版的车票预订系统。

2.1　车票预订系统用户及界面分析

本教材中所选车票预订系统主要应用于高等学校师生的乘车需求。在一些高校中，学生和教师需要乘车前往其他校区或地区，为了解决用户（学生、教师）长时间排队购票的问题，我们拟开发一个基于 C/S 结构的车票预订软件。

在专项实训过程中，我们重点关注学生应用相关技术实现预期项目的能力，有能力的学生可以在此基础上进行提升，如代码的优化、界面美化等。所以，在专项实训的实训要求中，对车票预订系统进行简化，关注主要的功能实现，以适应程序设计实训阶段学生的编程水平。

图 2-1　用户登录界面　　图 2-2　车票查票及预订界面

基于上述背景，我们首先对车票预订系统进行用户分析，本系统的用户主要包含普通用户（学生和教师）和系统管理员两种。其中，普通用户的操作权限包含登录、查询班车、预订车票三种，管理员的操作权限包含登录、开启服务器、关闭服务器三种。

本教材要求实现普通用户的登录、查票和订票的界面和功能；管理员对服务器进行管理的界面和功能，其界面效果可参考图2－1至图2－4。

图2－3　用户确认订票界面

图2－4　管理员功能界面

2.1.1　车票预订系统用例分析

根据以上分析，车票预订系统不同用户的用例描述如表2－1所示，管理员和普通用户两类用户的用例图如图2－5和图2－6所示。

图2－5　普通用户用例

图2－6　管理员用例

表2-1 用例描述

用　户	用　例
管理员（Admin）	登录、路线管理、车次管理、车票管理（后三个专项实训不做）
普通用户（User）	登录、车次查询、车票预订、车票信息卡、订单查询（专项实训不做）

2.1.2 车票预订系统数据分析

在车票预订系统中，需要存储的数据（实体）包含用户信息、班车信息、班车路线和车票信息，每类数据可通过一系列属性来表示，具体描述如下：

（1）用户（User）：用户编号（用户名）、姓名、密码、手机号码、专业、身份（学生、教师）。

（2）班车路线（Route）：路线编号（ID）、起点（starting）、终点（ending）。

（3）班车（Shuttle）：班车编号、路线编号、乘车日期、乘车时间、座位数、票价。

（4）车票信息（Ticket）：用户名、班车编号、乘车状态。

每类数据通过一个实体类进行描述，具体的实体类定义及实体之间的联系可通过图2-7所示内容描述，为方便管理，实体类全部放到info包中。

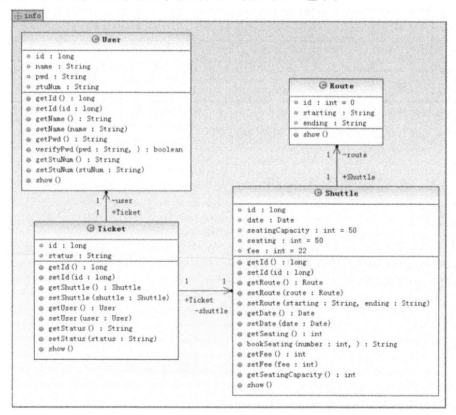

图2-7 实体类

2.2 专项实训增量划分及进度安排

车票预订系统实训项目采用增量式开发过程，根据采用 Java 技术的不同，分阶段实现所划分的增量，最终完成整个项目。通过增量式项目的开发，学生既可以巩固加深各个 Java 知识点的使用，又可以体会到一个完整项目的开发方法和过程，最终达到程序实训的真正目的。

开发阶段总划分：单机版软件开发阶段和客户机/服务器软件系统开发阶段。其中，单机版软件开发阶段包括界面开发、界面事件处理、数据库编程；客户机/服务器软件系统开发阶段包括采用 Socket 网络通信实现 C/S 结构的功能划分、采用多线程技术实现多用户并发访问功能。具体的实训阶段划分、每阶段需要的 Java 技术分布如表 2-2 所示。

表 2-2 增量开发阶段及步骤划分

	章节	界面编程	事件处理	数据库编程	数据流	网络通信	多线程
单机版	Ch3.1	√					
	Ch3.2		√				
	Ch4			√			
C/S 系统	Ch5				√	√	
	Ch6						√

注：为方便读者阅读，本文中采用如下标记方式：界面和界面元素均采用中文的中括号对标记，如【账号文本框】、【用户登录窗口】、【登录按钮】等。

实验 02 专项实训需求分析

实验目的
（1）掌握需求分析的过程与方法。
（2）掌握 StarUML 工具的使用。

专项实训题目
（1）学生请假管理系统。
（2）其他。

实验内容
（1）专项实训项目（学生请假管理系统）的功能性需求分析。
（2）专项实训增量开发计划。
（3）专项实训项目实体类设计。

实验安排
项目需求分析及增量式开发计划：3 学时。
实验报告
按照规定格式提交本次实验内容的结果。

第 3 章 GUI 编程及事件处理

本章以车票预订系统的界面设计为例，巩固和提高学生使用 Java 界面编程和事件处理技术，设计并实现专项实训项目的增量：界面和事件处理。

本章要求实现单机版车票预订系统的用户界面和事件处理，主要包括：

(1) 登录界面及其登录功能实现；
(2) 班车查询界面及其班车查询功能实现；
(3) 车票预订界面和车票预订功能实现；
(4) 实现以上各个界面之间的跳转。

下面我们将按照增量开发方法，逐步增量实现界面设计和界面事件处理两个用例。其中，界面设计的目的是为用户提供友好的图形化使用界面；事件处理提供了对应的用户操作机制，使界面之间"活动起来"。本章的重点是界面编程和事件处理，所以在事件处理中，我们通过显示固定数据（在程序中初始化数据）来显示操作结果。

3.1 车票预订系统的界面设计与实现

由于单机版车票预定系统的用例分为用户登录、班车查询和车票预订，所以，用户登录用例需要设计并实现用户登录界面；班车查询用例需要设计并实现班车查询、班车查询结果显示界面；车票预订需要设计并实现车票预订界面。

在项目实现中，我们把界面相关的类统一放入 gui 包，方便用户查看，界面类图如图 3-1 所示。

3.1.1 用户登录界面

1. 界面效果设计

为了实现用户登录功能，用户登录界面需要提供用户账号信息的输入信息框，其中用户账号需要采用明文显示，用户密码不能直接显示，除此之外，登录界面还应该提供用户确认登录的按钮。登录界面的参考效果如图 3-2 所示。

2. 实现登录界面需要的 Java 技术

首先，实现该【用户登录】窗口，我们可以采用 Java Swing 中的 JFrame 技术。由于登录界面的元素布局并不规则，不能采用 Java Swing 提供的布局管理器，故可以使用无布局管理器设置界面元素分布，实现该效果。

在界面中，【账号标签】和【密码标签】等提示信息可以使用标签 JLabel 实现，

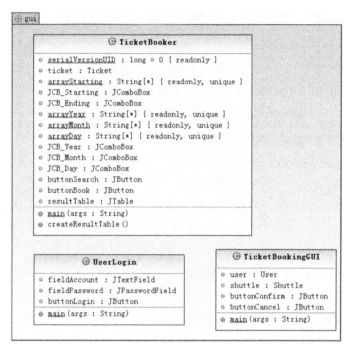

图 3-1 gui 包

图 3-2 登录界面

【账号文本框】可以采用 JTextField 实现,而【密码文本框】应该使用 JPasswordField 实现,【登录按钮】应该使用 JButton 实现。

3. 用户登录界面 UserLogin 类的设计

根据登录界面参考效果图和实现登录界面相关 Java 技术的分析,我们可以设计如图 3-3 所示的登录界面 UserLogin 类。实现登录界面的参考代码见【代码 3-1】。

```
┌─────────────────────────────────────┐
│           ⊖ UserLogin               │
├─────────────────────────────────────┤
│  ○ fieldAccount  : JTextField       │
│  ○ fieldPassword : JPasswordField   │
│  ○ buttonLogin   : JButton          │
│  ● main (args : String)             │
└─────────────────────────────────────┘
```

图 3-3 UserLogin 类

【代码 3-1】 UserLogin 类

```java
package gui;
import java.awt.BorderLayout;
import java.awt.Color;
import java.awt.GridLayout;
import javax.swing.BorderFactory;
import javax.swing.JButton;
import javax.swing.JFrame;
import javax.swing.JLabel;
import javax.swing.JPanel;
import javax.swing.JPasswordField;
import javax.swing.JTextField;
public class UserLogin extends JFrame {
private JTextField fieldAccount = new JTextField();
private JPasswordField fieldPassword = new JPasswordField();
private JButton buttonLogin = new JButton("登 录");

public static void main(String[] args) {
        new UserLogin();
}
public UserLogin() {
        this.setTitle("用户登录");
        this.setLocation(400, 300);
        this.setSize(300, 400);
        this.setLayout(null);
        this.setDefaultCloseOperation(EXIT_ON_CLOSE);
        this.setResizable(false);
        //用户账号
        JLabel labelAccount = new JLabel("账 号");
        labelAccount.setAlignmentY(CENTER_ALIGNMENT);
        labelAccount.setBounds(50, 100, 100, 30);
```

```java
        fieldAccount.setBounds(150,100,100,30);
        this.getContentPane().add(labelAccount);
        this.getContentPane().add(fieldAccount);
        //用户密码
        JLabel labelPwd = new JLabel("密码");
        labelPwd.setAlignmentY(CENTER_ALIGNMENT);
        labelPwd.setBounds(50,150,100,30);
        fieldPassword.setBounds(150,150,100,30);
        this.getContentPane().add(labelPwd);
        this.getContentPane().add(fieldPassword,
        BorderLayout.CENTER);
        //第6行为登录按钮
        buttonLogin.setBounds(75,250,150,30);
        this.getContentPane().add(buttonLogin);
        this.setVisible(true);
    }
}
```

在上述代码中，UserLogin 类通过继承 JFrame 类具备了 Java 窗体功能，该窗口中有 fieldAccount、fieldPassword 和 buttonLogin 三个属性，分别对应于用户的【账号文本框】、【密码文本框】和【登录按钮】。登录界面的创建由构造函数 UserLogin() 完成。为方便测试，在 UserLogin 类中添加了 main 函数，同时也提供了用户使用 UserLogin 类的例子。运行该功能，即可得到图 3-2 所示的用户登录界面。

3.1.2 班车查询及预订界面

1. 界面效果设计

在班车查询时，通过选择路线的起点和终点，选择乘车时间（年、月、日），然后单击【查询】按钮进行车票信息的查询。在查询结果中，选中一趟合适的车次，单击【预订】按钮即可进行订票操作。车票查询及预订界面的参考效果如图 3-4 所示。

2. 实现班车查询及预订界面需要的 Java 技术

类似于登录界面，使用 Java Swing 中的 JFrame 技术实现图 3-4 的窗口。在该班车查询及预订界面中，通过设置布局管理器实现不同的界面效果。整个界面划分为三部分：

（1）最上端的面板实现班车查询功能，通过 FlowLayout 布局管理器实现，在该布局上添加班车查询条件各元素。【路线】等提示信息标签通过 JLabel 实现，【路线起点】、【路线终点】和【班车时间】通过 JComboBox 实现，【查询】通过 JButton 实现。

（2）中间面板用来显示班车查询的结果，通过 BorderLayout 布局管理器实现，在该布局上使用 JTable 显示查询得到的班车信息。

（3）最下端面板主要用来实现预订功能，在查询到的班车信息中，选择一条合适信息后，单击【预订】按钮以实现车票预订功能。通过在 BorderLayout 布局管理器中添

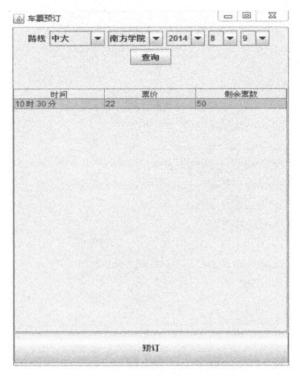

图 3-4 班车查询及预订界面

加 JButton 实现。

3. 班车查询及预订 TicketBooker 类的设计

根据界面效果和实现界面相关 Java 技术的分析,我们可以设计如图 3-5 所示的班车查询及预订界面 TicketBooker 类。实现班车查询及预订界面的参考代码见【代码 3-2】。

图 3-5 TicketBooker 类

【代码 3-2】 TicketBooker 类

```java
package gui;
import java.awt.BorderLayout;
import java.awt.FlowLayout;
import java.util.Calendar;
import java.util.Date;
import info.Ticket;
import info.User;
import javax.swing.JButton;
import javax.swing.JComboBox;
import javax.swing.JFrame;
import javax.swing.JLabel;
import javax.swing.JPanel;
import javax.swing.JScrollPane;
import javax.swing.JTable;
import javax.swing.table.JTableHeader;
import javax.swing.table.TableColumn;
import javax.swing.table.TableModel;
public class TicketBooker extends JFrame{
private static final long serialVersionUID = 7285459016493447805L;
    public static void main(String[] args) {
            User user = new User();
            user.setId(1001);
            user.setName("user");
            user.setStuNum("1001");
            new TicketBooker(user);
    }
    private Ticket ticket = new Ticket();
    //路线 - 起点、终点,使用下拉菜单
    private static final String arrayStarting[] = {"中大","南方学院"};
    private JComboBox JCB_Starting = new JComboBox(arrayStarting);
    private JComboBox JCB_Ending = new JComboBox(arrayStarting);
    //private JButton buttonCalendar = new JButton("日期");
    //private Calendar calendar = Calendar.getInstance();
    private static final String arrayYear[] = {"2014","2015","2016"};
    private static final String arrayMonth[] = {"1","2","3","4","5","6","7","8","9","10","11","12"};
    private static final String arrayDay[] = {"1","2","3","4","5","6","7","8","9","10","11","12","13","14","15","16","17","18","19","20","21","22","23","24","25","26","27","28","29","30","31"};
    private JComboBox JCB_Year = new JComboBox(arrayYear);
    private JComboBox JCB_Month = new JComboBox(arrayMonth);
```

```java
    private JComboBox JCB_Day = new JComboBox(arrayDay);
    private JButton buttonSearch = new JButton("查询");private JButton button-
Book = new JButton("预订");
    private JTable resultTable = null;
    public TicketBooker(User user){
        this.setTitle("车票预订");
        this.setLocation(400,300);
        this.setSize(400,600);
        this.setDefaultCloseOperation(EXIT_ON_CLOSE);
        this.setLayout(null);
        //设置布局,页面分上中下三部分:
        //上面为选项,依次为:JLabel Route, JComboBox JCB_Starting JCB_Ending,
JButton Calendar 选择,JButton 查询
        //中间为查寻结果,采用JTable:依次为时间,票价,现有车票数,选中任何一行后即
可单击【预订】按钮
        //下面为JButton 预订,单击后显示预订结果
        JPanel upper = new JPanel();
        JPanel middle = new JPanel();
        JPanel bottom = new JPanel();
        upper.setLocation(0,0);
        upper.setSize(400,100);
        middle.setLocation(0,100);
        middle.setSize(400,400);
        bottom.setLocation(0,500);
        bottom.setSize(400,50);
        //在上面的面板添加元素
        upper.setLayout(new FlowLayout());
        upper.add(new JLabel("路线"));
        JCB_Starting.setToolTipText("起点");
        upper.add(JCB_Starting);
        JCB_Ending.setToolTipText("终点");
        upper.add(JCB_Ending);
        JCB_Year.setToolTipText("年");
        upper.add(JCB_Year);
        JCB_Month.setToolTipText("月");
        upper.add(JCB_Month);
        JCB_Day.setToolTipText("日");
        upper.add(JCB_Day);
        //设置当前日期
        Calendar calendar = Calendar.getInstance();
        Date date = calendar.getTime();
        JCB_Year.setSelectedIndex(date.getYear() + 1900 - Integer. par-
```

```
seInt(arrayYear[0]));
            JCB_Month.setSelectedIndex(date.getMonth());
            JCB_Day.setSelectedIndex(date.getDate()-1);
            upper.add(new JLabel());
            //upper.add(buttonCalendar);
            upper.add(buttonSearch);
            this.add(upper);
            //在中间的 Panel 上添加元素
            middle.setLayout(new BorderLayout());
            //resultTable = 初始化
            createResultTable();
            JScrollPane scrollPane = new JScrollPane(resultTable);
            middle.add(scrollPane);
            this.add(middle);
            //在下面的 Panel 上添加元素
            bottom.setLayout(new BorderLayout());
            bottom.add(buttonBook, BorderLayout.CENTER);
            this.add(bottom);
            //显示界面
            this.setVisible(true);
    }
    private void createResultTable(){
            if (resultTable != null) return;

            Object[][] data = {};
            //JTable:依次为时间,票价,现有车票数,选中任何一行后即可单击【预订】按钮
            String[] name = {"时间", "票价", "剩余票数"};
            resultTable = new JTable(data, name);
        }
    }
```

3.1.3 用户确认订票信息界面

1. 界面效果设计

用户单击【预订】按钮后,为使用户确认订票信息,会弹出用户确认订票信息界面,在该界面将显示用户所订车票的乘车信息,用户可以单击【确认】按钮完成订票操作,也可以单击【取消】按钮将车票取消,如图3-6所示。

2. 实现该界面需要的 Java 技术

在该界面中,共显示四行信息:【用户信息】、【路线(起点和终点)】、【乘车时间】、【确认】和【取消】按钮,分别使用 GridLayout 布局管理器,并在各布局管理器上添加所需的界面元素。其中【用户信息】、【路线(起点和终点)】、【乘车时间】通过 JLabel 实现,【确认】和【取消】按钮通过 JButton 实现。

图3-6 用户确认订票信息

3. 用户确认订票信息界面TicketBookingGUI类的设计

根据界面效果和实现界面相关Java技术的分析,我们可以设计如图3-7所示的确认订票信息界面类TicketBookingGUI。实现用户确认订票信息界面的参考代码见【代码3-3】。

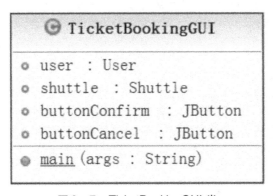

图3-7 TicketBookingGUI类

【代码3-3】TicketBookingGUI类

```java
package gui;
import java.awt.FlowLayout;
import java.awt.GridLayout;
import java.util.Calendar;
import java.util.Date;
import info.Route;
import info.Shuttle;
import info.User;
import javax.swing.JButton;
import javax.swing.JFrame;
import javax.swing.JLabel;
import javax.swing.JPanel;
public class TicketBookingGUI extends JFrame {
    private User user = null;
    private Shuttle shuttle = null;
```

```java
    private JButton buttonConfirm = new JButton("确认");
    private JButton buttonCancel = new JButton("取消");
    public TicketBookingGUI(User user, Shuttle shuttle){
        this.user = user;
        this.shuttle = shuttle;
        this.setTitle("车票信息");
        this.setLocation(500,300);
        this.setSize(300,160);
        this.setResizable(false);
        this.setDefaultCloseOperation(DISPOSE_ON_CLOSE);
        this.setLayout(new GridLayout(4,1,0,0));
        JPanel panel = new JPanel();
        //第1行:用户名,学号
        panel.setLayout(new GridLayout(1,4,0,0));
        panel.add(new JLabel("用户名"));
        panel.add(new JLabel(user.getName()));
        panel.add(new JLabel("学号"));
        panel.add(new JLabel(user.getStuNum()));
        this.add(panel);
        //第2行:路线:起点,终点
        panel = new JPanel();
        panel.setLayout(new GridLayout(1,4,0,0));
        panel.add(new JLabel("起点"));
        panel.add(new JLabel(shuttle.getRoute().starting));
        panel.add(new JLabel("终点"));
        panel.add(new JLabel(shuttle.getRoute().ending));
        this.add(panel);
        //第3行:日期
        panel = new JPanel();
        panel.setLayout(new GridLayout(1,2,0,0));
        panel.add(new JLabel("时间"));
        Date date = shuttle.getDate();
        panel.add(new JLabel("" + date.getYear() + "年" + date.getMonth() +
"月" + date.getDate() + "日" + date.getHours() + "时" + date.getMinutes() +
"分"));
        this.add(panel);
        //第4行:确认按钮,取消按钮
        panel = new JPanel();
        panel.setLayout(new GridLayout(1,2,0,0));
        panel.add(buttonConfirm);
        panel.add(buttonCancel);
        this.add(panel);
```

```java
        this.setVisible(true);
}
    public static void main(String[] args) {
            User user = new User();
            user.setId(100001);
            user.setName("张三");
            user.setStuNum(Integer.toString(10001));
            Route route = new Route();
            route.id = 1;
            route.starting = "中大";
            route.ending = "南方学院";
            Calendar calendar = Calendar.getInstance();
            calendar.set(2014, 9, 1, 10, 30); //2009年9月1日10时30分
            Date date = calendar.getTime();
            Shuttle shuttle = new Shuttle(50);
            shuttle.setRoute(route);
            shuttle.setDate(date);
            shuttle.setFee(22);
            shuttle.setId(1);
            new TicketBookingGUI(user, shuttle);
    }
}
```

3.2 事件处理（以 ActionListener 为例）

在 3.1 中，我们只是设计了车票预订系统的主要窗口界面，并没有实现相应的功能，如登录、班车查询、预订车票等。添加事件处理后，用户界面即可响应用户的操作，界面之间可以实现跳转，从而使系统"真正工作"起来。

本部分以 ActionListener 为例，讲解简单的事件处理方法，读者还可以采用其他事件处理方法，本教材不再详述。

3.2.1 用户登录事件处理

1. 用户登录事件处理过程

实现用户登录功能，可采取以下处理流程：

（1）在图 3-2 中，用户在【账号文本框】输入账号，在【密码文本框】中输入密码后，单击【登录】按钮；

（2）用户登录界面开始处理用户的登录请求：从【账号文本框】中读出账号，【密码文本框】中读出密码，并与已有账号（例如：账号"user1"，密码"user1"）进行匹配，实现账号验证；

（3）账号验证通过，弹出图 3-4 所示的【班车查询及预订】窗口；

（4）如果账号验证失败，则提示用户登录失败，如弹出【登录失败】消息框，并

显示"账号/密码错误，请重新登录"等字样，如图 3-8 所示，读者也可以采取其他方式实现登录失败提示，如把账号和密码内容设为红色，或者增加一个文本标签显示登录失败信息等，本书不再详述。

图 3-8　登录失败消息框

2. 用户登录事件处理相关技术

GUI 需要由事件驱动，Java 事件处理机制是一种事件处理框架，其设计目的是把 GUI 交互动作（单击、菜单选择等）转变为调用相关的事件处理程序进行处理。对事件进行处理是通过事件监听器实现的。事件监听器对应的接口一般位于包 java.awt.event 和 javax.swing.event 中。文本框、按钮、菜单项、密码框和单选按钮都可以触发 ActionEvent 事件，Java 中对能触发 ActionEvent 事件的组件的使用方法是，为其增加 addActionListener（ActionListener listen）方法，将实现 ActionListener 接口的类的实例注册为事件源的监听器。

ActionListener 接口在 java.awt.event 包中，该接口中只有一个方法：Public void actionPerformed（ActionEvent e），当事件源触发 ActionEvent 事件后，监听器调用实现的接口中的方法 actionPerformed（ActionEvent e），以对发生的事件作出处理。

在用户登录事件中，对用户账号的验证方法可采用字符串比较方法，如 String 的 equals（）方法进行验证。

【用户登录失败】信息提示窗口，可以使用 showMessageDialog（）方法，如 JOptionPane.showMessageDialog（this,"账号或密码错误"）。

用户登录信息验证成功后，会实现窗体的跳转，弹出【班车查询及预订】窗口，同时关闭【用户登录】窗口。新窗体弹出方式可以通过创建一个新的 TicketBooker 类来实现，关闭【用户登录】窗口可以通过调用 setVisible（flase）设置当前 UserLogin 对象为不可见，或通过调用 dispose（）方法销毁【用户登录】窗口。

3. 更新的 UserLogin 类

在上述分析的基础上，我们更新图 3-3 所示的 UserLogin 类，使之通过扩展 ActionListener 接口，为【登录】按钮 buttonLogin 增加事件处理监听器。本例中增加了用户验证私有方法 verifyAccount（），并重写 ActionListener 接口的 actionPerformed（）方法。新的 UserLogin 类结构如图 3-9 所示。

```
┌─────────────────────────────────────────────────┐
│                  UserLogin                       │
├─────────────────────────────────────────────────┤
│ ○ fieldAccount : JTextField                     │
│ ○ fieldPassword : JPasswordField                │
│ ○ buttonLogin : JButton                         │
│ ● main (args : String)                          │
│ ● actionPerformed (e : ActionEvent)             │
│ ● verifyAccount (account : String, password : String, ) : User │
└─────────────────────────────────────────────────┘
```

图 3-9　更新的 UserLogin 类

4. 用户登录事件处理流程

根据用户登录事件处理过程和更新的 UserLogin 类，绘制用户登录的时序图如图 3-10 所示。

图 3-10　用户登录时序

5. 用户登录事件处理代码实现

【代码 3-4】用户登录事件处理

```java
public class UserLogin extends JFrame implements ActionListener{
    ...
    public UserLogin() {
        ...
        //第 6 行为登录按钮
        buttonLogin.setBounds(75,250,150,30);
        this.getContentPane().add(buttonLogin);
        buttonLogin.addActionListener(this);
        this.setVisible(true);
    }
```

```java
@Override
public void actionPerformed(ActionEvent e) {
        if (e.getSource() = = buttonLogin){
                String account = fieldAccount.getText();
                String password = new String(fieldPassword.getPassword
                ());
                User user = verifyAccount(account, password);
                if( user ! = null){
                        //弹出车票预订窗口
                        new TicketBooker(user);
                        //关闭当前窗口
                        this.dispose();
                }
                else{
                        JOptionPane.showMessageDialog(this,"账号或密
                        码错误");
                        fieldPassword.setText(null);
                }
        }
}

private User verifyAccount(String account, String password) {
        //这里固定用户名和密码为10001/10001
if(account.equals("10001") && password.equals("10001")){
        User user = new User();
        user.setId(10001);
        user.setName("10001");
        user.setStuNum("1001");
        return user;
    }
    return null;
 }
}
```

3.2.2 班车查询及订票事件处理

1. 班车查询及订票事件的处理过程

实现班车查询功能，可采取的处理流程为：

(1) 在图 3-4 中【组合框】中，用户分别选择班车的起点、终点、发车时间（包括年、月、日）后，单击【查询】按钮；

(2) 班车查询界面处理用户的查询请求，从【组合框】中依次读取起点、终点和发车时间，并与已有的班车信息进行匹配，并将匹配正确的班车信息以列表的形式显示

在界面中;

(3) 每次查询都会将上次查询结果清除,并显示本次查询结果;

(4) 在查询结果中选择一条班车信息后,单击【预订】按钮实现订票事件;

(5) 预订成功,弹出图3-6所示的【用户确认订票信息】窗口。

2. 班车查询及订票事件处理相关技术

班车查询及订票使用的GUI事件处理技术和用户登录事件处理类似,在该查询事件中,通过shuttleSearch()方法查询是否有班车,并通过updateResultTable(searchResult)方法将查询到的班车信息添加到JTable中。

在订票事件中,需要选中一条班车信息才能单击【预订】按钮,通过getSelectedRow()方法选中一条班车信息,showMessageDialog()函数用来判断是否选中班车。

用户订票成功后,会实现窗体的跳转,弹出【确认订票信息】窗口,同时关闭【班车查询及订票】窗口。新窗体弹出方式可以通过创建一个新的TicketBookingGUI类来实现,关闭【班车查询及订票】窗口可以通过调用setVisible(flase)设置当前TicketBooker对象为不可见,或通过调用dispose()方法销毁【班车查询及订票】窗口。

3. 更新的TicketBooker类

在上述分析的基础上,我们更新图3-5所示的TicketBooker类,使之通过扩展ActionListener接口,为【查询】按钮buttonSearch和【预订】按钮buttonBook增加事件处理监听器。本例中增加了查询班车私有方法shuttleSearch()和更新班车信息私有方法updateResultTable(searchResult),并重写ActionListener接口的actionPerformed()方法。新的TicketBooker类如图3-11所示。

图3-11 更新的TicketBooker类

4. 班车查询及订票事件处理流程

根据班车查询及订票事件处理过程和更新的 TicketBooker 类,绘制班车查询及订票的时序图如图 3–12 所示。

图 3–12 班车查询及订票时序

5. 班车查询及订票事件处理代码实现

【代码 3–5】班车查询及订票事件处理

```
package gui;
public class TicketBooker extends JFrame implements ActionListener {
    ...
    public TicketBooker(User user) {
        ...
        //设置布局,页面分上中下三部分:
        //上面为选项,依次为:JLabel Route, JComboBox JCB_Starting JCB_Ending, JButton
        //Calendar 选择, JButton 查询
        //中间为查寻结果,采用 JTable:依次为时间,票价,现有车票数,选中任何一行后即可单击预订按钮
        //下面为 JButton 预订,单击后显示预订结果
        ...
        //在上面的面板添加元素
        ...
        upper.add(buttonSearch);
        buttonSearch.addActionListener(this);
        this.add(upper);
        //在中间的 Panel 上添加元素
        ...
```

```java
        //在下面的Panel上添加元素
        bottom.setLayout(new BorderLayout());
        bottom.add(buttonBook, BorderLayout.CENTER);
        buttonBook.addActionListener(this);
        this.add(bottom);
        //显示界面
        this.setVisible(true);
    }
    private void createResultTable() {
        if (tableResult != null)
            return;
        Object[][] data = {};
        //JTable:依次为时间,票价,现有车票数,选中任何一行后即可单击预订按钮
        String[] name = { "时间", "票价", "剩余票数" };
        tableModelDefault = new DefaultTableModel(data, name);
        tableResult = new JTable(tableModelDefault);
        //设置为单选
        tableResult.setSelectionMode(ListSelectionModel.SINGLE_SELECTION);
    }
    private void updateResultTable(Vector<Shuttle> searchResult) {
        if (tableResult == null || tableModelDefault == null)
            return;
        //清空查询结果
        tableModelDefault.setRowCount(0);
    for (int i = 1; i <= searchResult.size(); i++) {
        Shuttle shuttle = searchResult.get(i-1);
        Date date = shuttle.getDate();
        //增加第i行
        String time = "" + date.getHours() + "时" + date.getMinutes() + "分";
        Object data[] = { time, shuttle.getFee(), shuttle.getSeating() };
        tableModelDefault.addRow(data);
        }
    }
    private Vector<Shuttle> shuttleSearch() {
            String starting = (String) JCB_Starting.getSelectedItem();
            String ending = (String) JCB_Ending.getSelectedItem();
            int year = Integer.parseInt((String) JCB_Year.getSelectedItem());
            int month = Integer.parseInt((String) JCB_Month.getSelectedItem());
            int day = Integer.parseInt((String) JCB_Date.getSelectedItem());
            //查询是否有班车
            //这里固定车次,每天都有相同的时刻的车次
```

```java
            //中大 → 南方学院:10:30
            //南方学院 → 中大:12:30
            Calendar calendar = Calendar.getInstance();
            int hour = 10;
            int minute = 30;
            //清空上次查询记录
            searchResult.clear();
            Shuttle shuttle = new Shuttle(50);
            shuttle.setFee(22);
            shuttle.setId(1);
            shuttle.setRoute(starting, ending);
            if (starting.equals("中大") && ending.equals("南方学院")) {
                hour = 10;
                minute = 30;
    } else if (starting.equals("南方学院") && ending.equals("中大")) {
                hour = 12;
                minute = 30;
        }
        calendar.set(year, month, day, hour, minute);
        Date date = calendar.getTime();
        shuttle.setDate(date);
        searchResult.add(shuttle);
        return searchResult;
}
@Override
public void actionPerformed(ActionEvent e) {
        if (e.getSource() == buttonSearch) {
            Vector<Shuttle> searchResult = shuttleSearch();
            updateResultTable(searchResult);
        } else if (e.getSource() == buttonBook) {
            //tableResult 与 searchResult 的内容是按顺序一一对应的
            int selectedRow = tableResult.getSelectedRow();
            //检查是否选中某个班车
            if (selectedRow < 0) {
                JOptionPane.showMessageDialog(this, "请选择车次(单选)");
            } else {
                new TicketBookingGUI(user, searchResult.get(selectedRow));
            }
        }
    }
}
public static void main(String[] args) {
        User user = new User();
```

```
        user.setId(1001);
        user.setName("user");
        user.setStuNum("1001");
        new TicketBooker(user);
    }
}
```

3.2.3 确认订票信息事件处理

1. 确认订票信息事件处理过程

实现确认订票信息，可采取以下处理流程：

(1) 在上一步操作中，用户单击【预订】按钮订票成功后，弹出图3-6所示的【确认订票信息】窗口；

(2)【确认订票信息】窗口中列出了用户所订车票的详细信息，包括：用户名、学号、起点、终点、发车时间等；

(3) 用户可对所订车票进行确认，单击【确认】按钮完成订票操作，也可以单击【取消】按钮将车票取消。

2. 确认订票信息事件处理实现相关技术

在确认订票信息事件处理中，实现技术类似前述步骤，重写 ActionListener 接口的 actionPerformed() 方法，实现事件触发。

3. 更新的 TicketBookingGUI 类

在上述分析的基础上，我们更新图3-7所示的 TicketBookingGUI 类，使之通过扩展 ActionListener 接口，为【确认】按钮 buttonConfirm 和【取消】按钮 buttonCancel 增加事件处理监听器。本例中并重写 ActionListener 接口的 actionPerformed() 方法。新的 TicketBookingGUI 类如图3-13所示。

图3-13 更新的 TicketBookingGUI 类

4. 确认订票信息事件处理流程

根据班车查询及订票事件处理过程和更新的 TicketBookingGUI 类，绘制确认订票信息的时序图如图3-14所示。

5. 确认订票信息事件处理代码实现

【代码3-6】确认订票信息事件处理
```
package gui;
public class TicketBookingGUI extends JFrame implements ActionListener{
```

图 3-14 确认订票信息时序

```
...
public TicketBookingGUI(User user, Shuttle shuttle){
    ...
    /* 界面效果
     * 第1行:用户名,学号
     * 第2行:路线:起点,终点
     * 第3行:日期
     */
    ...
    //第4行:确认按钮,取消按钮
    panel = new JPanel();
    panel.setLayout(new GridLayout(1,2,0,0));
    panel.add(buttonConfirm);
    panel.add(buttonCancel);
    buttonConfirm.addActionListener(this);
    buttonCancel.addActionListener(this);
    this.add(panel);
    this.setVisible(true);
}
@Override
public void actionPerformed(ActionEvent e) {
    if(e.getSource() == buttonConfirm){
        System.out.println(">>>>>>>>>>>>预订成功<<<<<<<<<<<<<<");
        ticket.show();
    }
    else if(e.getSource() == buttonCancel){
        System.out.println(">>>>>>>>>>>>取消成功<<<<<<<<<<<<<<");
    }
```

```java
            dispose();
    }

    public static void main(String[] args) {
        User user = new User();
        user.setId(100001);
        user.setName("张三");
        user.setStuNum(Integer.toString(10001));

        Route route = new Route();
        route.id = 1;
        route.starting = "中大";
        route.ending = "南方学院";
        Calendar calendar = Calendar.getInstance();
        calendar.set(2014, 9, 1, 10, 30); //2009年9月1日10时30分
        Date date = calendar.getTime();
        Shuttle shuttle = new Shuttle(50);
        shuttle.setRoute(route);
        shuttle.setDate(date);
        shuttle.setFee(22);
        shuttle.setId(1);
        new TicketBookingGUI(user, shuttle);
    }
}
```

实验 03　GUI 编程

实验目的

（1）掌握 Java 图形界面编程的基本流程。
（2）掌握 GUI 编程的基本方法。
（3）掌握 JButton 等常见组件的使用方法。
（4）掌握事件监听、监听注册和事件处理的概念、基本方法。
（5）提高编程的逻辑思维能力和程序排错能力。

实验内容

在实验 02 的基础上，完成以下内容：
（1）学生请假管理系统的登录界面设计。
（2）学生请假管理系统的主界面设计。
（3）学生请假管理系统的各分界面设计。

实验安排

（1）学生请假管理系统的登录界面、主界面：3 学时。
（2）各分界面设计：3 学时。

实验报告

按照规定格式提交本次实验内容的结果。

实验 04 事件处理

实验目的

（1）掌握 JButton 事件监听、监听注册和事件处理的简单方法；

（2）掌握字符串 String 的简单应用；

（3）掌握基本界面元素的使用方法；

（4）掌握 JTable 的使用方法。

实验内容

在实验 03 的基础上，完成各界面中以下各功能的事件触发：

（1）登录按钮的事件触发。

（2）学生请假按钮的事件触发。

（3）辅导员修改假条、查询假条的事件触发。

（4）教师查询假条的事件触发。

（5）其他事件触发。

实验安排

（1）实验内容（1）、（2）：3 学时；

（2）实验内容（3）至（5）：3 学时。

实验报告

按照规定格式提交本次实验内容的结果。

第4章 数据库编程

本章以车票预订系统的数据库设计为例,巩固和提高学生使用Java的数据库连接及处理技术,设计并实现专项实训项目的增量:数据库编程。

本章要求实现单机版车票预订系统的数据库编程,主要包括:

(1) 班车预订系统的数据库设计;
(2) 登录功能的数据库实现;
(3) 班车查询功能的数据库实现;
(4) 班车预订功能的数据库实现。

下面我们将按照增量开发方法,逐步增量实现班车预订系统的数据库编程,以使数据实现永久存取。

4.1 数据库设计

4.1.1 E-R图

根据第2章车票预订系统需求分析得到的用例图及用例描述,设计并得到车票预订系统的E-R图,如图4-1所示。

图4-1 车票预订系统E-R

4.1.2 数据库表设计

由图4-1的E-R图,设计车票预订系统的数据库(CPYD)由用户表、班车信息表和订票信息表三个表组成,如表4-1所示。表4-2至4-4详细定义了各表的结构及约束。

表4-1 车票预订系统数据库

表 名	描 述
用户表 userlist	用于存储用户基本信息
班车信息表 shuttlelist	用于存储不同班车的信息
订票信息表 ticketlist	用于存储用户的订票信息

表4-2 用户信息表 userlist

序号	字段	名称	数据类型	P	U	F	I	C	备注
1	id	学生 id	BIGINT	√					NOT NULL
2	name	学生姓名	VARCHAR(50)						NOT NULL
3	stuNum	学号	VARCHAR(30)						NOT NULL
4	pwd	密码	VARCHAR(20)						NOT NULL

表4-3 班车信息表 shuttlelist

序号	字段	名称	数据类型	P	U	F	I	C	备注
1	id	班车 id	BIGINT	√					NOT NULL
2	s_starting	始发站	VARCHAR(50)						NOT NULL
3	s_ending	终点站	VARCHAR(50)						NOT NULL
4	s_date	发车日期	DATE						NOT NULL
5	s_time	发车时间	TIME						NOT NULL
6	capacity	容纳人数	SMALLINT						default 50
7	seating	剩余座位	SMALLINT						default 50

表4-4 订票信息表 ticketlist

序号	字段	名称	数据类型	P	U	F	I	C	备注
1	id	车票 id	BIGINT	√					NOT NULL
2	shuttle_id	班车 id	BIGINT						NOT NULL
3	user_id	用户 id	BIGINT						NOT NULL
4	status	乘车状态	VARCHAR(20)						NOT NULL

4.1.3 数据库脚本

本教材选用的数据库管理系统为 MySQL。在 MySQL 下，输入车票预订系统数据库创建的 T-SQL 脚本，即可完成车票预订系统 CPYD 的数据库创建，CPYD 系统的数据库脚本见【代码 4-1】。

【代码 4-1】CPYD 数据库脚本

```sql
create database if not exists CPYD;
use CPYD;

CREATE TABLE userlist
(
id    BIGINT not null,
name    VARCHAR(50) not null,
stuNum    VARCHAR(30) not null,
pwd    VARCHAR(20),

PRIMARY KEY (id)
) DEFAULT CHARSET = utf8;

CREATE TABLE shuttlelist
(
id    BIGINT not null auto_increment,
s_starting VARCHAR(50) not null,
s_ending VARCHAR(30) not null,
s_date    DATE,
s_time    TIME,
capacity    smallint default 50,
seating    smallint default 50 check(seating >= 0),
PRIMARY KEY (id)
) DEFAULT CHARSET = utf8;

CREATE TABLE ticketlist
(
id    BIGINT not null auto_increment,
shuttle_id    BIGINT not null,
user_id    BIGINT not null,
status    VARCHAR(20),

PRIMARY KEY (id)
) DEFAULT CHARSET = utf8;
```

4.2 JDBC 配置

车票预订系统使用 JDBC 数据源配置方式，具体的配置步骤如下：
1. 加载 JDBC 驱动

（1）在 MyEclipse 集成开发环境下，选中项目 CPYD 后单击鼠标右键，依次选择 Build Path→Configure Build Path，如图 4-2 所示。

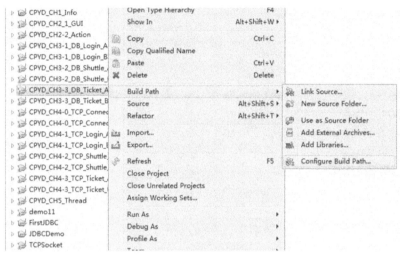

图 4-2　Build Path（1）

（2）在图 4-3 中依次选择 Libraries→Add External JARs 命令，在弹出的 JAR Selection 文件选择窗口中选择要添加到项目的 JDBC 驱动（例如：mysql-connector-java-5.0.8-bin.jar），单击【确定】按钮即可看到 JDBC 驱动已经加载进图 4-3 所示窗口。

图 4-3　Build Path（2）

2. 建立连接

定义用于进行数据库操作的类文件 CPYDDatabase.java，建立与 MySQL 下 CPYD 数据库的访问连接，具体代码见【代码 4-2】。

【代码 4-2】 CPYDDatabase 类

```java
package database;
import info.User;
import java.sql.Connection;
import java.sql.DriverManager;
import java.sql.ResultSet;
import java.sql.SQLException;
import java.sql.Statement;
public class CPYDDatabase {
  private static Connection conn = null;
  public CPYDDatabase() {
  }
  //取得数据库连接
private static Connection getConnection() {
        if (conn ! = null) {
                return conn;
        }
        String driver_MySQL = "com.mysql.jdbc.Driver";
        String url = "jdbc:mysql://localhost:3306/CPYD";
        String account_MySQL = "root";
        String password_MySQL = "root";
        try {
                Class.forName(driver_MySQL);
                conn = DriverManager.getConnection(url, account_MySQL, password_MySQL);
        } catch (Exception e) {
                e.printStackTrace();
                System.out.println("创建数据库连接失败!");
        }
        return conn;
    }
}
```

4.3 登录功能的数据库实现

4.3.1 用户登录时序图

在 3.2.1 节的基础上，增加访问数据库操作，修改图 3-10 时序图，绘制具有数据库操作的用户登录时序图,如图4-4所示。

图4-4 用户登录的数据库实现

4.3.2 功能实现

在用户登录界面,根据用户的登录信息(用户名和密码)与数据库中存储的用户信息进行匹配,如果匹配成功则进入查询界面,否则提示用户登录失败。由用户登录时序图可知,用户登录包含两个操作步骤:①在用户登录界面提取用户信息;②将提取的用户信息与数据库中的数据进行匹配。所以,需要修改第3章相应的类,增加访问数据库功能。

1. 数据库操作脚本

在CPYDDatabase类文件中增加查询数据库的脚本,见【代码4-3】。

【代码4-3】用户信息查询数据库脚本

```
…
private static String toSqlString(String str){
        return new String("'" + str + "'");
}
//account is stuNum
public static User userQquery(String accountName){
        User user = null;
         String sql = "select * from userlist where name = " + toSqlString(accountName);
    try{
        Statement stmt = getConnection().createStatement();
        ResultSet rs = stmt.executeQuery(sql);
        if(rs.next() == true){
            user = new User();
            user.setId(rs.getLong(1));
```

```
                user.setName(rs.getString(2));
                user.setStuNum(rs.getString(3));
                user.setPwd(rs.getString(4));
            }
        } catch (SQLException sqle) {
            System.out.println("查询数据出现异常:" + sqle.getMessage());
        }
    }
    return user;
}
```

2. 用户身份验证

在 UserLogin 类中修改用户身份验证 verifyAccount() 方法，通过调用查询数据库脚本的 userQquery() 方法来实现对用户身份的识别，具体见【代码 4-4】。

【代码 4-4】用户身份验证

```
private User verifyAccount(String account, String password) {
    //这里固定用户名和密码为 10001/10001
    User user = CPYDDatabase.userQquery(account);
    if (user == null){
        System.out.println("用户不存在");
    } else if (!user.verifyPwd(password)) {
        System.out.println("密码错误");
        user = null;
    }
    //else,验证通过

    return user;
}
```

4.4 班车查询功能的数据库实现

4.4.1 班车查询功能时序图

在 3.2.2 节基础上，增加访问数据库操作，修改图 3-12 时序图，绘制具有数据库操作的班车查询时序图，如图 4-5 所示。

4.4.2 功能实现

1. 数据库操作脚本

在 CPYDDatabase 类文件中增加班车查询功能的数据库操作代码，具体见【代码 4-5】。

【代码 4-5】班车查询数据库脚本

```
...
/**
```

图 4-5 班车查询的数据库实现

```
 * @ param starting
 * @ param ending
 * @ param year
 * @ param month
 * @ param day
 * @ return 为 null 表示没有相应车次,否则查询到
 */
public static Vector < Shuttle > shuttleQquery(String starting, String ending,
            int year, int month, int day) {
    Vector < Shuttle > result = new Vector < Shuttle >();
    Shuttle shuttle = null;
    Calendar calendar = Calendar.getInstance();
    Date date = null;
    String sql = "select id, s_time, capacity, seating from shuttlelist where"
        + "s_starting = "
        + toSqlString(starting)
        + "AND"
        + "s_ending = "
        + toSqlString(ending)
        + "AND"
        + "s_date = "
        + toSqlString(new String("" + year + "-" + month + "-" + day));
    System.out.println(sql);
```

```
        try {
          Statement stmt = getConnection().createStatement();
          ResultSet rs = stmt.executeQuery(sql);
          while (rs.next() = = true) {
          long id = rs.getLong(1);
          String str = rs.getString(2); //TIME, e.g.'12:30:00'
          int capacity = rs.getShort(3);
          int seating = rs.getShort(4);
          int hourOfDay = Integer.parseInt(str.substring(0, 2));
          int minute = Integer.parseInt(str.substring(3, 5));
          calendar.set(year, month, day, hourOfDay, minute);
          date = calendar.getTime();
          shuttle = new Shuttle(capacity, seating);
          shuttle.setId(id); // ID
          shuttle.setRoute(starting, ending); //s_starting, s_starting
          shuttle.setDate(date);
          result.add(shuttle);
          }
        } catch (SQLException sqle) {
           System.out.println ("[shuttleQquery]查询数据出现异常: " + sqle.getMessage());
        }
        return result;
    }
```

2. 班车查询

在 TicketBooker 类中修改查询班车信息的 shuttleSearch() 方法，通过调用 shuttleQquery() 方法实现对数据库的查询操作，具体见【代码 4 - 6】。

【代码 4 - 6】班车查询

```
  ...
  private Vector < Shuttle > shuttleSearch() {
        String starting = (String) JCB_Starting.getSelectedItem();
        String ending = (String) JCB_Ending.getSelectedItem();
        int year = Integer.parseInt((String) JCB_Year.getSelectedItem());
         int month = Integer.parseInt((String) JCB_Month.getSelectedItem());
        int day = Integer.parseInt((String) JCB_Date.getSelectedItem());
        //查询是否有班车
        searchResult.clear();
        Vector < Shuttle > searchResult = CPYDDatabase.shuttleQquery(starting, ending, year, month, day);
        return searchResult;
    }
```

```
@Override
public void actionPerformed(ActionEvent e) {
    if (e.getSource() == buttonSearch) {
        Vector<Shuttle> searchResult = shuttleSearch();
        updateResultTable(searchResult);
    }
    else if (e.getSource() == buttonBook) {
        //tableResult 与 searchResult 的内容是按顺序一一对应的
        int selectedRow = tableResult.getSelectedRow();
        //检查是否选中某个班车
        if (selectedRow < 0) {
            JOptionPane.showMessageDialog(this,"请选择车次(单选)");
        }
        else {
            new TicketBookingGUI(user, searchResult.get(selectedRow));
        }
    }
}
```

4.5 班车预订功能的数据库实现

4.5.1 班车预订功能时序图

在3.2.3节基础上，增加访问数据库操作，修改图3-14时序图，绘制具有数据库操作的班车预订时序图，如图4-6所示。

图4-6 班车预订的数据库实现

4.5.2 功能实现

1. 数据库操作脚本

在 CPYDDatabase 类文件中增加班车预定的数据库操作代码,具体见【代码 4-7】。

【代码 4-7】班车预订数据库脚本

```
…
/**
 * @param shuttleId
 * @param userId
 * @return true 表示预订成功,false 表示预订失败
 */
public static synchronized boolean ticketBook(long shuttleId, long userId){
    String sql = null;
    try{
    //更新 shuttlelist
    Statement stmt = getConnection().createStatement();
    sql = "update shuttlelist set seating = seating - 1 where id = " + shuttleId;
    stmt.executeUpdate(sql);
    sql = "insert into ticketlist(shuttle_id, user_id, status) values (" + shuttleId + ", " + userId + ",'已预订');";
    stmt.executeUpdate(sql);

    return true;
    } catch (SQLException sqle) {
    System.out.println("[ticketBook]查询数据出现异常: " + sqle.getMessage());
    }
    return false;
}
```

2. 班车预订

在 TicketBookingGUI 类中修改 actionPerformed() 方法,通过 handleTicketBook() 方法调用 ticketBook() 方法实现对班车的预订功能,见代码【4-8】。

【代码 4-8】班车预订

```
…
public void actionPerformed(ActionEvent e){
    if(e.getSource() == buttonConfirm){
        //向数据库中增加记录,更新班次中车票数量
        if(handleTicketBook(ticket.getShuttle().getId(), ticket.getUser().getId())){
            JOptionPane.showMessageDialog(this, "订票成功");
        }
```

```
        else{
          JOptionPane.showMessageDialog(this,"订票失败");
        }
      }
      else if(e.getSource() = = buttonCancel){
        System.out.println(">>>>>>>>>>>>>取消成功<<<<<<<<<<<<<<");
      }

      dispose();
    }

    private boolean handleTicketBook(long shuttleId, long userId) {
      return CPYDDatabase.ticketBook(shuttleId, userId);
    }
```

实验05　数据库编程

实验目的

（1）掌握数据库设计的步骤。
（2）掌握 MySQL 的使用方法。
（3）掌握 T‑SQL 的基本语法。
（4）掌握 JDBC 配置步骤。
（5）掌握访问数据库数据的实现方法。

实验内容

（1）学生请假管理系统数据库设计。
（2）实现学生请假管理系统的 JDBC 配置，建立数据库连接。
（3）MySQL 下学生请假管理系统的数据库操作方法。
（4）实现学生请假管理系统数据的访问。

实验安排

（1）数据库设计、JDBC 配置及数据库连接：3 学时。
（2）数据库访问：3 学时。

实验报告

按照规定格式提交本次实验内容的结果。

第 5 章 网络通信编程

本章以车票预订系统的网络通信为例,学习和掌握通过 Socket 在客户端和服务端之间建立连接渠道来进行数据的传输和通信的能力,设计并实现专项实训项目的增量:网络通信编程。

本章要求实现简单的网络版车票预定系统,主要包括:
(1) TCP 对象数据流通信功能的搭建与设计;
(2) 客户端与服务器之间通过对象数据流实现登录功能;
(3) 客户端与服务器之间通过对象数据流实现班车查询;
(4) 客户端与服务器之间通过对象数据流实现车票预订功能。

下面我们将按照增量开发方法,结合前面的内容逐步增量实现网络通信编程的用例。本章的重点是 TCP 对象数据流的交互。在搭建网络中,我们通过获取数据库信息来显示操作结果。

5.1 网络通信技术分析

所谓网络编程就是在两个或两个以上的设备(如计算机)之间传输数据,而 Java 的网络编程大致可以分为两类:一种是通过 URL 来获取网络上的资源;另一种是通过 Socket 在客户端与服务器之间建立连接通道来进行数据的传输和通信。

5.1.1 网络编程的基本知识

1. 通信协议

网络通信的核心是协议。协议是指进程之间为完成任务在交换信息的过程中所使用的一系列规则和规范。进程之间只要遵循相同的网络协议,即使在不同的语言环境下编写也可以进行信息交互。TCP/IP(Transmission Control Protocol/Internet Protocol)是 Internet 的通信协议,称为传输控制/网际协议。它是一组由 TCP 协议和 IP 协议、UDP(User Datagram Protocol)协议等协议组成的。TCP/IP 提供了可靠的网络通信环境。

2. 网络地址

网络上每一个主机都有唯一的地址作为标识,这个地址叫做 IP 地址。IPv4 使用了 32 位地址,实际表现为一个四点格式的十进制数据,例如:202.103.141.9。每个数字代表一个 8 位二进制数。

3. 端口

计算机通过端口与外界进行交流，端口的范围是 0～65536，其中 0～1023 是一些与服务有关的端口。

4. 套接字

套接字（Socket）是用来描叙 IP 地址和端口的通信端点。使用套接字可以实现两个程序之间的连接和通信。

5. URL

URL（Uniform Resource Location）称为"统一资源定位器"，是对可以从互联网上得到的资源的位置和访问方法的一种简洁的表示，是互联网上标准资源的地址。URL 地址的格式如下：

协议：// 主机名［：端口号］/路径/［；参数］［? 动态网页传输参数］#资源片段

可以参照下面一个例子：http：//www.nfu.edu.cn

5.1.2 URL 通信方式

网络上的网页通常是以 URL 方式进行访问的。Java.net 包提供了 URL 类和 URLConnection 类来从远程站点获取信息的方法。一般步骤有 3 步。

1. 创建 URL 对象

我们可以通过下面的构造方法来初始化一个 URL 对象。

（1）public URL（String spec）。

通过一个表示 URL 地址的字符串可以构造一个 URL 对象。

例：URL myURL = new URL（" http：//www.nfu.edu.cn/"）。

（2）public URL（URL context, String spec）。

通过基 URL 和相对 URL 构造一个 URL 对象。

例：URL myURL = new URL（" http：//www.nfu.edu.cn/"）；
URL indexURL = new URL（myURL," index.html"）。

（3）public URL（String protocol, String host, String file）。

通过协议 protocol，主机名 host 和文件路径 file 构造一个 URL 对象。

（4）public URL（String protocol, String host, int port, String file）。

通过协议 protocol、主机名 host、端口号 port 和文件路径 file 构造一个 URL 对象。

在构造方法中如果制定的协议不合法，会抛出 MalformedURLException 异常，因此生成 URL 对象时，我们必须要对这一例外进行处理，通常是用 try – catch 语句进行捕获。

2. 读取页面信息

当我们得到一个 URL 对象后，就可以通过 URL 的方法 openStream（）读取指定的网页资源，并返回类型为 InputStream 的对象。在这里可能抛出 IOException 异常。

【代码 5 – 1】URL 例子

```
URL url = new URL("http://www.nfu.edu.cn");
```

```
BufferedReader br = new BufferedReader(new InputStreamReader(
url.openStream()));
String line = null;
while(null! = (line = br.readLine())){
System.out.println(line);
}
br.close();
```

3. 创建一个到 URL 的连接

通过 URLConnection 类，程序可以同 URL 资源进行交互，具体步骤如下：

（1）使用 openConnection（）方法创建连接对象。

URL url = new URL（"http: //www. nfu. edu. cn"）;

URLConnection con = url. openConnection（）;

（2）向服务器写数据。

建立数据流：PrintStream out = new PrintStream（con. getOutputStream（））;

写数据：out. println（String data）;

（3）向服务器读数据。

建立输入数据流：

InputStreamReader in = new InputStreamReader（con. getInputStream（））;

BufferedReader inb = new BufferedRead（in）;

向服务器读取数据：in. readLine（）;

5.1.3 套接字通信方式

1. TCP 套接字通信

首先，要通过 TCP 套接字进行通信就需要使用到 java. net. * 和 java. io. * 里面的多个类，前者提供实现网络应用程序的类包括：Socket、ServerSocket；后者提供通过数据流、序列化和文件系统提供系统输入和输出的方法，如 InputStream、DataInputStream 等。

根据 TCP Socket 进行通信的过程，建立 TCP 所需要的 Java 技术包括客户端和服务端。

客户端：

（1）提供用于描述的 IP 地址和端口来建立 TCP Socket。

Socket socket = new Socket（130. 100. 1. 23，3021）;

（2）发送数据流。

创建发送数据流：OutputStream os = socket. getOutputStream（）;

创建发送对象数据流：ObjectOutputStream oos = new ObjectOutputStream（os）;

通过数据流发送对象：oos. writeObject（…）;

（3）接收数据流。

创建接收数据流：InputStream is = socket. getInputStream（）;

创建接收对象数据流：ObjectInputStream ois = new ObjectInputStream（is）；

通过数据流接收对象：ois. ReadObject（）；

（4）关闭数据流和 Socket。

关闭数据流：socket. shutdownOutput（）；

socket. shutdownInput（）；

关闭 Socket：socket. close（）；

服务端：

（1）提供端口号创建 ServerSocket：

ServerSocket serverSocket = new ServerSocket（serverPort）。

（2）接收到客户端访问：

Socket socket = serverSocket. accept（）；

（3）接收数据流。

创建接收数据流：InputStream is = socket. getInputStream（）；

创建接收对象数据流：ObjectInputStream ois = new ObjectInputStream（is）；

通过数据流接收对象：ois. ReadObject（…）；

（4）发送数据流。

创建发送数据流：OutputStream os = socket. getOutputStream（）；

创建发送对象数据流：ObjectOutputStream oos = new ObjectOutputStream（os）；

通过数据流发送对象：oos. writeObject（…）；

（5）关闭数据流和 Socket。

关闭数据流：socket. shutdownOutput（）；

socket. shutdownInput（）；

关闭 Socket：socket. close（）；

2. UDP 套接字通信

要通过 UDP 套接字进行通信就需要使用到 java. net 包里面的 DatagramSocket 类和 DatagramPacket 类。前者用于在程序中建立传送数据报的通信连接，后者则用来表示一个数据报。使用 UDP 进行通信的时候，服务器需要有一个线程不停地监听客户端的请求。

客户端：

根据 UDP Socket 进行通信的过程，建立 UDP 客户端编程涉及的步骤也是 4 个部分：建立连接、发送数据、接收数据和关闭连接。

（1）建立 UDP Socket。

端口号可以指定，下面的例子是系统本身随机分配的一个本地计算机的未用端口号：

DatagramSocket ds = new DatagramSocket（）；

（2）发送数据报：ds. send（DatagramPacket packet）；

（3）接收数据报。

构造一个数据缓冲数组：byte［］data = new byte［1024］；

构造一个 DatagramPacket 数据包对象：

DatagramPacket receiveDp = new DatagramPacket（data，data.length）；

接收数据报：ds.receive（receiveDp）；

（4）关闭连接：ds.close（）；

服务端：

（1）提供端口号创建服务器监听 Socket：

DatagramSocket ds = new DatagramSocket（10010）；

（2）接收数据：ds.receive（）；

（3）关闭服务器：ds.close（）；

5.2　TCP 对象数据流通信功能的设计与搭建

在项目实现中，我们使用套接字来实现建立 TCP 对象数据流通信机制。但是由于对象数据流只能传递经过序列化的对象，所以我们要先对已建立的对象进行序列化的处理。

在项目实现中，建立 TCP 对象数据流通信机制，需要搭建客户端 ClientDummy 类和服务器端 ServerAdmin 类的对象数据流接受平台，如图 5 – 1 所示。

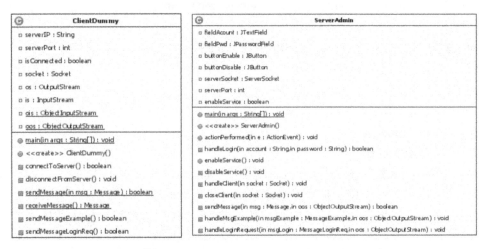

(a) ClientDummy 类图　　　　　　(b) ServerAdmin 类图

图 5 – 1　ClientDummy 和 ServerAdmin 类

其次，在项目里面，我们建立了 Message 类用以封装信息并进行信息传递，信息类图如图 5 – 2 所示。

5.2.1　对象序列化的实现

1. 对象序列化处理过程

Java 里面网络通信是通过流（Stream）来实现，序列化就是将内存中的类或者对象（我们写的类都是存储在内存中的）变成可以存储到存储媒介中的流。在项目实现中，

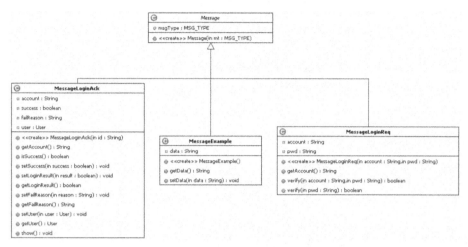

图 5-2 Message 类图

我们把在网络通信中需要传递的对象进行序列化处理。具体步骤为：
（1）选择需要序列化的对象；
（2）对对象进行序列化处理。

2. 对象序列化需要的 Java 技术

首先，一个对象要有序列化特性就必须实现接口 java.io.Serializable。在实现了这个接口后，对象才可以转化为数据类（Stream）并在通信功能中进行传递。实现 Serializable 接口的方式为：class Message implements Serializable。

3. 在项目中实现序列化的功能

在项目中，客户端和服务器需要交互的对象有用户信息（User）、路线信息（Route）、班车信息（Shuttle）和车票信息（Ticket）以及用以发送信息的信息类（Message）。

【代码 5-2】实现该功能的代码修改如下所示：

```
//用户信息(User)类
package info;
import java.io.Serializable;
public class User implements Serializable{
…
}


//路线信息(Route)类
package info;
import java.io.Serializable;
public class Route implements Serializable{
…
}
```

```
//班车信息(Shuttle)类
package info;
import java.io.Serializable;
public class Shuttle implements Serializable {
…
}
//车票信息(Ticket)类
package info;
import java.io.Serializable;
public class Ticket implements Serializable {
…
}
```

【代码5-3】信息类的创建以及序列化处理

```
package messages;
import java.io.Serializable;
    public abstract class Message implements Serializable {
        public enum MSG_TYPE {
MSG_UNKOWN, MSG_EXAMPLE;
}
public MSG_TYPE msgType = MSG_TYPE.MSG_UNKOWN;
public Message(MSG_TYPE mt) {
msgType = mt;
}
}
```

上述代码中，类 User、Route、Shuttle、Ticket 和 Messae 都是通过导入 java.io.Serializable 接口并实现这个接口来达到对象序列化功能。

5.2.2 对象数据流通信功能的搭建

1. TCP 套接字（Socket）通信的基本原理和过程

为了支持 TCP/IP 面向链接的网络程序开发，java.net 提供了 Socket 类和 ServerSocket 类。其中，ServerSocket 用于服务器端程序，Socket 类则用于编写客户端程序。客户端和服务器之间通过输入输出流进行通信。

在上述过程中，需要注意的是客户端的端口号要与服务器的端口号一致，服务器才能接收到客户端发送的数据。在数据的通信过程中，客户端或服务器要先发送数据，另一端才可以接收到相应的数据。

2. TCP 套接字处理流程

通过 Java 相关技术分析，在项目里面，ClientDummy 类通过创建布尔值型 connectToServer() 来访问服务器 serverAdmin。如果访问成功，ClientDummy 类发送 sendMessageExample() 方法；否则 ClientDummy 类会在控制台上输出失败信息。最后，ClientDummy 类通过方法 disconnectFromServer() 关闭套接字和所有数据流，绘制班车查询及订票的时序图如图 5-3 所示。

图 5-3 项目网络通信时序

3. 车票订位通信网络代码实现

【代码 5-4】客户端 ClientDummy 类

```java
package client;
import java.io.IOException;
import java.io.InputStream;
import java.io.ObjectInputStream;
import java.io.ObjectOutputStream;
import java.io.OutputStream;
import java.net.Socket;
import messages.Message;
import messages.MessageExample;
public class ClientDummy {
    public static void main(String[] args) {
        new ClientDummy();
    }
    //假设 Server IP 为本机 IP
    private String serverIP = "127.0.0.1";
    private int serverPort = 54321;
    //记录是否已经建立了到服务器的 TCP socket 连接
    private boolean isConnected = false;
    private Socket socket = null;
    private OutputStream os = null;
    private InputStream is = null;
```

```java
        private static ObjectInputStream ois = null;
        private static ObjectOutputStream oos = null;
        public ClientDummy() {
            if (connectToServer()) {
                System.out.println("sendMessageExample: " + sendMessageExample());
            } else {
                System.out.println("Failed to connect to server" + serverIP + "/"
                        + serverIP);
            }
            try {
                Thread.sleep(1000);  //等1000毫秒后退出
            } catch (InterruptedException e) {
                e.printStackTrace();
            }
            disconnectFromServer();
        }

        private boolean connectToServer() {
            //Socket尚未初始化,在端口打开连接,取出对象输入输出流
            try {
                socket = new Socket(serverIP, serverPort);
                socket.setSoTimeout(3000);  //如果服务器没有反映,尝试3000毫秒
                os = socket.getOutputStream();
                is = socket.getInputStream();
                oos = new ObjectOutputStream(os);
                ois = new ObjectInputStream(is);
                isConnected = true;
            } catch (IOException e) {
                isConnected = false;
            } finally {
            }
            return isConnected;
        }
        private void disconnectFromServer() {
            if (socket != null) {
                try {
                    socket.shutdownOutput();
                    socket.shutdownInput();
                    socket.close();
                    socket = null;
                    oos = null;
```

```java
            ois = null;
        } catch (IOException e) {
            // TODO Auto-generated catch block
            e.printStackTrace();
        }
    }
}
private static synchronized boolean sendMessage(Message msg) {
    try {
        oos.writeObject(msg);
    } catch (IOException e) {
        e.printStackTrace();
        return false;
    }
    return true;
}
private static synchronized Message receiveMessage() {
    try {
        return (Message) ois.readObject();
    } catch (ClassNotFoundException e) {
        e.printStackTrace();
    } catch (IOException e) {
        e.printStackTrace();
    }
    return null;
}

private boolean sendMessageExample() {
    MessageExample msgSnt = new MessageExample();
    msgSnt.setData("Hello");

    if (sendMessage(msgSnt)) {
        // 接收服务器返回的消息并显示出来
        MessageExample msgRev = (MessageExample) receiveMessage();
        if (msgRev != null) {
          System.out.println("Message Received: " + msgRev.getData());
            return true;
        }
    }
    return false;
  }
}
```

【代码5-5】服务器 serverAdmin 类

```java
package server;
import java.awt.GridLayout;
import java.awt.event.ActionEvent;
import java.awt.event.ActionListener;
import java.awt.event.WindowAdapter;
import java.awt.event.WindowEvent;
import java.io.EOFException;
import java.io.IOException;
import java.io.InputStream;
import java.io.ObjectInputStream;
import java.io.ObjectOutputStream;
import java.io.OutputStream;
import java.net.ServerSocket;
import java.net.Socket;
import java.net.SocketException;
import java.net.SocketTimeoutException;
import javax.swing.JButton;
import javax.swing.JFrame;
import javax.swing.JLabel;
import javax.swing.JOptionPane;
import javax.swing.JPasswordField;
import javax.swing.JTextField;
import messages.Message;
import messages.MessageExample;

public class ServerAdmin extends JFrame implements ActionListener {

    public static void main(String[] args) {
        new ServerAdmin();
    }
    //界面元素
    private JTextField fieldAcount = new JTextField(20);
    private JPasswordField fieldPwd = new JPasswordField(20);
    private JButton buttonEnable = new JButton("开启服务");
    private JButton buttonDisable = new JButton("关闭服务");
    //网络服务信息
    private ServerSocket serverSocket = null;
    private int serverPort = 54321;
    //服务开关
    private boolean enableService = false;
    public ServerAdmin() {
```

```java
        this.setTitle("Server");
        this.setLocation(300,200);
        this.setSize(300,120);
        this.setResizable(false);
        this.addWindowListener(new WindowAdapter() {
            public void windowClosing(WindowEvent e) {
                disableService();
                System.exit(0);
            }
        });
        //设置页面布局
        this.setLayout(new GridLayout(3,2,0,0));
        this.add(new JLabel("帐 号"));
        this.add(fieldAcount);
        this.add(new JLabel("密 码"));
        this.add(fieldPwd);
        buttonDisable.setEnabled(false);
        this.add(buttonEnable);
        this.add(buttonDisable);
        buttonEnable.addActionListener(this);
        buttonDisable.addActionListener(this);

        this.setVisible(true);
        //启动服务器 TCP 端口和服务
        enableService();
        //用户几点关闭服务按钮,系统退出,关闭 serverSocket
        disableService();
}
@Override
public void actionPerformed(ActionEvent e) {
    if (e.getSource() == buttonEnable) {
        String account = fieldAcount.getText();
        String password = new String(fieldPwd.getPassword());
        //判断用户名和密码是否正确
        //登陆成功的条件:用户账号存在,密码正确, 管理员权限
if (handleLogin(account, password)) {

            //禁止登录按钮,使能退出按钮
            buttonEnable.setEnabled(false);
            buttonDisable.setEnabled(true);

            //启动 ServerSocket
```

```java
            enableService = true;
        } else {
            JOptionPane.showMessageDialog(this, "Invalid User");
            fieldPwd.setText("");
        }
    } else if (e.getSource() == buttonDisable) {
        //系统退出服务,系统退出
        enableService = false;
        disableService();
        System.exit(0);
    }
}
private boolean handleLogin(String account, String password) {
    if (account.equals("admin") && password.equals("admin"))
        return true;
    return false;
}
public void enableService() {
    //等待启动服务
    while (! enableService) {
        try {
            Thread.sleep(500);
        } catch (InterruptedException e) {
            e.printStackTrace();
        } finally {
        }
    }
    //开放服务端口
    try {
        serverSocket = new ServerSocket(serverPort);
    } catch (IOException e) {
        e.printStackTrace();
        return;
    }
    //使用while可以允许client端多次连接,但由于没有使用线程,同时只能有一个客户端连接服务器
    while (enableService) {
        //监听端口请求,等待连接
        try {
            //通过端口建立连接
            Socket socket = serverSocket.accept();
            //收到客户端连接,处理用户请求
```

```java
            System.out.println("Request received");
            handleClient(socket);

            //客户端退出
            closeClient(socket);
            socket = null;
        } catch (SocketTimeoutException e) {
            //do nothing
        } catch (IOException e) {
            e.printStackTrace();
        } finally { //停止服务
        }
    }
}
private void disableService() {
    if (serverSocket != null) {
        try {
            serverSocket.close();
        } catch (IOException ioe) {
            ioe.printStackTrace();
        }
        serverSocket = null;
    }
}
private void handleClient(Socket socket) {
    OutputStream os = null;
    ObjectOutputStream oos = null;
    InputStream is = null;
    ObjectInputStream ois = null;
    try {
        //创建对象数据流
        os = socket.getOutputStream();
        is = socket.getInputStream();
        oos = new ObjectOutputStream(os);
        ois = new ObjectInputStream(is);
    } catch (IOException e) {
        System.out.println(e.getMessage());
        return;
    }
    //处理来自客户端的消息对象
    while (enableService) {
        //从对象输入数据流中读取数据对象
```

```java
            Message msgReceived = null;
            try {
                msgReceived = (Message) ois.readObject();
            } catch (ClassNotFoundException e) {
                System.out.println(e.getMessage());
                continue;
            } catch (IOException e) { // Socket 出错,停止对次 Socket 的处理
                System.out.println(e.getMessage());
                return;
            }
            // 按消息类型处理
            switch (msgReceived.msgType) {
            case MSG_EXAMPLE:
                handleMsgExample((MessageExample) msgReceived, oos);
                break;
            default:
                System.out.println("Uknown message received: "
                        + msgReceived.msgType);
            }
        }
    }
    private void closeClient(Socket socket) {
        if (socket != null) {
            try {
                socket.shutdownOutput();
                socket.shutdownInput();
                socket.close();
            } catch (IOException e) {
                e.printStackTrace();
            }
        }
    }
    private boolean sendMessage(Message msg, ObjectOutputStream oos) {
        try {
            oos.writeObject(msg);
        } catch (IOException e) {
            e.printStackTrace();
            return false;
        }
        return true;
    }
    private void handleMsgExample(MessageExample msgExample,
```

```
        ObjectOutputStream oos) {
    String data = msgExample.getData();
    System.out.println(data);
    msgExample.setData("Example Message Received: " + data);
    sendMessage(msgExample, oos);
    }
}
```

在上述代码中，serverAdmin 类通过继承 JFrame 类具备了窗口功能，窗口界面设计如图 5-4 所示。在输入了正确的用户名和密码后，用户可以点击【开启服务】按钮调用函数 enableService（） 来开启服务器，点击【关闭服务】按钮来调用函数 disableService（） 来关闭服务器。

图 5-4　ServerAdmin 窗口界面

5.3　TCP 对象数据流通信功能的实现

5.3.1　TCP 用户登录功能实现

1. 用户登录功能实现过程

利用对象数据流实现用户登录功能，可采取以下流程：

（1）创建一个用户登录验证请求消息类和一个登录验证结果消息类。

（2）客户端在【登录】按钮的事件处理过程中，把用户名和密码保存在用户登录验证请求消息对象中，通过输出对象数据流，发给服务器。

（3）服务器端：服务器从输入对象数据流中收到用户登录验证请求消息，根据消息对象中的用户名，从数据库中检查该用户是否存在。

（4）服务器端：用户存在，比较密码是否正确，如果用户存在并且密码正确，创建登录验证结果消息对象，设置登录状态为成功，把该消息发给客户端。

（5）服务器端：用户不存在或者密码不正确，设置登录状态为失败，把该消息发给客户端。

（6）客户端：根据接收到的登录验证结果，显示相应界面。

2. 实现用户登录功能需要的 Java 技术

使用对象数据流的用户登录功能，需要使用到套接字、接收数据流和发送数据流等技术。在前面搭建好数据通信环境的基础上，修改客户端在【登录】按钮的 ActionListener 事件处理上，把用户名和密码从 JTextFiled 里面获取并包装到 MessageLoginReq 类。

在该用户登录功能的实现中，客户端利用对象数据流通过 MessageLoginReq 类发送用户登录账号和密码到服务器验证,服务端通过 MessageLoginAck 返回用户登录结果以及原因到客户端。

3. 用户登录功能的设计

根据实现登录功能相关 Java 技术的分析,我们可以创建一个用户登录验证请求消息类 MessageLoginReq 和一个登录验证结果消息类 MessageLoginAck,类图设计如图 5-2,代码参见【代码 5-6】和【代码 5-7】,并通过客户端 UserLogin 类调用 MessageLoginReq 和 MessageLoginAck 来实现用户登录功能。

【代码 5-6】MessageLoginReq 类

```java
package messages;
import java.io.Serializable;
public class MessageLoginReq extends Message implements Serializable{
    private String account = null;
    private String pwd = null;
    public MessageLoginReq(String account, String pwd){
        super(MSG_TYPE.MSG_LOGIN_REQ);
        this.account = account;
        this.pwd = pwd;
    }
    public String getAccount(){
        return account;
    }
    public boolean verify(String account, String pwd){
        if (this.account.equals(account) && this.pwd.equals(pwd))
            return true;

        return false;
    }
    public boolean verify(String pwd){
        if (pwd.equals(pwd))
            return true;
        return false;
    }
}
```

【代码 5-7】MessageLoginAck 类

```java
package messages;
import info.User;
import java.io.Serializable;
public class MessageLoginAck extends Message implements Serializable{
    private String account = null;
    private boolean success = false;
    private String failReason = null;
```

```java
    private User user = null;
    public MessageLoginAck(String id) {
        super(MSG_TYPE.MSG_LOGIN_ACK);
        account = id;
    }
    public String getAccount() {
        return account;
    }
    public boolean isSuccess() {
        return success;
    }
    public void setSuccess(boolean success) {
        this.success = success;
    }
    public void setLoginResult(boolean result) {
        success = result;
    }
    public boolean getLoginResult() {
        return success;
    }
    public void setFailReason(String reason) {
        failReason = reason;
    }
    public String getFailReason() {
        return failReason;
    }
    public void setUser(User user) {
        this.user = user;
    }
    public User getUser() {
        return user;
    }
    public void show() {
        System.out
                .println(">>>>>>>>>>>>>>>>>>> MessageLoginAck <<<<<<<<<<<<<<<<<<<");
        System.out.println("account: " + account);
        System.out.println("success: " + success);
        System.out.println("failReason: " + failReason);
        user.show();
    }
}
```

在上述代码中，MessageLoginReq 类和 MessageLoginAck 类都继承了 Message 父类的方法，并实现了 Serializable 接口。

5.3.2　TCP 班车查询及订票功能实现

1. 用户班车查询功能实现过程

利用对象数据流实现班车查询功能，可采取的实现流程为 4 个步骤。

（1）创建一个班车查询消息类和一个班车查询结果消息类。

（2）客户端：把查询条件存入（封装进）一个班车查询消息类对象，把该对象发给服务端。

（3）服务器端：收到一个班车查询消息类对象，根据查询条件，从数据库中查询班车信息，把结果存入一个班车查询结果消息类对象，发给客户端。

（4）客户端：收到一个班车查询结果消息类对象，把结果显示在界面上。

利用对象数据流实现班车订票功能，可采取的实现流程为 4 个步骤。

（1）创建一个订票修改请求消息类和一个订票修改结果消息类。

（2）客户端：把班车 ID 和用户 ID 以及订票信息存入（封装进）一个订票修改请求消息类对象，把该对象发给服务端。

（3）服务器端：收到一个订票修改请求消息类对象，根据对象中信息存入数据库，更新班车信息，把数据库更新结果存入一个成绩修改结果消息类对象，发给客户端。

（4）客户端：收到一个订票修改请求结果消息类对象，把结果显示在界面上。

2. 实现班车查询及订票功能需要的相关技术

班车查询的对象数据流技术和用户登录的对象数据流处理过程相似。在该查询事件中需要使用到班车查询消息类 MessageShuttleQuery 的对象把用户查询的条件包装并从客户端发送到服务器；服务器在接收到 MessageShuttleQuery 的对象发送的信息后，根据查询条件，从数据库中查询班车信息，把结果存入一个班车查询结果消息类 MessageShuttleQueryAck 的对象，并发给客户端。

与班车查询的对象数据流技术相似，班车订票的对象数据流需要使用到订票修改请求消息类 MessageTicketBook 的对象把班车 ID 和用户 ID 以及订票信息封装并从客户端发送到服务器；服务器在接收到 MessageTicketBook 的对象发送的信息后，根据对象中信息存入数据库，更新班车信息，把数据库更新结果存入一个成绩修改结果消息类 MessageTicketBookAck 的对象，并发给客户端。

3. 班车查询及订票功能的设计

根据实现班车查询以及订票功能的相关 Java 技术的分析，我们通过创建班车查询消息类 MessageShuttleQuery 和一个班车查询结果消息类 MessageShuttleQueryAck，类图设计如图 5-5 所示，代码参见【代码 5-8】和【代码 5-9】，来实现班车查询功能；通过创建订票修改请求消息类 MessageTicketBook 和一个成绩修改结果消息类 MessageTicketBookAck 来实现订票功能类图设计如图 5-6 所示，代码参见【代码 5-10】和【代码 5-11】。

图 5-5　MessageShuttleQuery 类和 MessageShuttleQueryAck 类

【代码 5-8】班车查询消息类 MessageShuttleQuery

```java
package messages;
import java.io.Serializable;
public class MessageShuttleQuery extends Message implements Serializable{
    private String starting = null;
    private String ending = null;
    private int year = 2014;
    private int month = 1;
    private int date = 1;
    public MessageShuttleQuery(String starting, String ending, int year, int month, int date) {
        super(MSG_TYPE.MSG_SHUTTLE_QUERY);
        this.starting = starting;
        this.ending = ending;
        this.year = year;
        this.month = month;
        this.date = date;
    }
    public String getStarting() { return starting;}
    public String getEnding() { return ending; }
    public int getYear() { return year;}
    public int getMonth() { return month; }
    public int getDate() { return date;}
}
```

【代码 5-9】班车查询结果消息类 MessageShuttleQueryAck

```java
package messages;
import info.Shuttle;
import java.io.Serializable;
import java.util.Vector;
public class MessageShuttleQueryAck extends Message implements Serializable{
```

```java
    Vector<Shuttle> shuttles = null;
    public Vector<Shuttle> getShuttles(){ return shuttles;}
    public void setShuttles(Vector<Shuttle> shuttles) {this.shuttles = shuttles;}
    public MessageShuttleQueryAck()
    {super(MSG_TYPE.MSG_SHUTTLE_QUERY_ACK);}
    public void show(){
        if(shuttles ! = null){
            for(Shuttle shuttle : shuttles){
                shuttle.show();
            }
        }
    }
}
```

图 5-6 MessageTicketBook 类和 MessageTicketBookAck 类

【代码 5-10】订票修改请求消息类 MessageTicketBook

```java
package messages;
import java.io.Serializable;
public class MessageTicketBook extends Message implements Serializable{
    private long shuttleId = 0;
    private long userId = 0;
    public MessageTicketBook(long shuttleId, long userId) {
        super(MSG_TYPE.MSG_TICKET_BOOK_REQ);
        this.shuttleId = shuttleId;
        this.userId = userId;
    }
    public long getShuttleId() {
        return shuttleId;
    }
```

```
    public long getUserId() {
        return userId;
    }
}
```

【代码 5-11】 成绩修改结果消息类 MessageTicketBookAck

```
package messages;
import java.io.Serializable;
public class MessageTicketBookAck extends Message implements Serializable{
    private long shuttleId = 0;
    private long userId = 0;
    private boolean success = false;
    public MessageTicketBookAck(long shuttleId, long userId) {
        super(MSG_TYPE.MSG_TICKET_BOOK_ACK);
        this.shuttleId = shuttleId;
        this.userId = userId;
    }
    public long getShuttleId() {
        return shuttleId;
    }
    public long getUserId() {
        return userId;
    }
    public boolean isSuccess() {
        return success;
    }
    public void setSuccess(boolean success) {
        this.success = success;
    }
}
```

实验 06　网络通信编程

实验目的

(1) 掌握 Java 对象序列化的基本方法。
(2) 掌握 Java 的网络通信功能。
(3) 掌握 Socket 通信的概念。
(4) 掌握网络通信传递过程中对象数据流的构造。
(5) 提高编程的逻辑思维能力和程序排错能力。

实验内容

在实验 04 的基础上，完成以下内容：
(1) 学生请假管理系统的客户端与服务器的搭建。

（2）学生请假管理系统的消息父类的设计与实现。
（3）学生请假管理系统的各功能的消息类的设计与实现。

实验安排

（1）学生请假管理系统的客户端与服务器的搭建：3学时。
（2）学生管理系统的实现登录验证方法的对象数据流的设计与实现：3学时。
（3）学生管理系统其他对象数据流的设计与实现：3学时。

实验报告

按照规定格式提交本次实验内容的结果。

第6章 多线程编程

本章以车票预订系统的多线程编程为例,学习和掌握通过定义多线程的子类来使得服务器同时支持与多个客户端进行双向通信的能力,设计并实现专项实训项目的增量:多线程。

本章要求实现多线程的车票预订系统,主要包括:
(1) 车票预订系统实现线程的创建;
(2) 车票预订系统可以处理多个用户的访问;
(3) 车票预订系统的线程同步问题。

下面我们将按照增量开发方法,逐步增量实现车票预订系统的多线程处理能力。

6.1 车票预订系统的多线程创建

线程的概念来源于计算机的操作系统的进程的概念。进程是一个程序关于某个数据集的一次运行。而线程是一个运行在一个程序的上下文内的轻量级进程。简单来说线程(thread)则是进程中的一个单个的顺序控制流;而多线程是指在单个的程序内可以同时运行多个不同的线程完成不同的任务,如图6-1所示。

图6-1 单线程与多线程

在 Java 语言中,线程的创建有两种方法。它们分别是:
(1) 继承 Thread 类并覆盖它的 run() 方法;
(2) 实现 Runnable 接口并实现它的 run() 方法。

在项目实现中,我们在前一章所建立的服务器的基础上,选择了继承 Thread 类的方法,并覆盖该类中 run() 的方法来创建服务器的线程并通过调用线程中需要的方法来实现操作,服务器线程类 ServerThread 类图如图 6-2 所示。

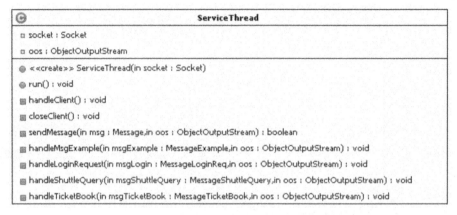

图 6-2　ServiceThread 类

1. 线程创建实现过程

通过定义一个 Thread 类的服务器子类来实现线程的创建,可采用以下流程:

(1) 定义一个继承 Thread 类的服务器子类;

(2) 在子类中重写 Thread 类的 run() 方法,从而定义线程所需完成的工作;

(3) 在 run() 方法调用方法 handleClient() 处理用户事件,在用户事件结束后通过方法 closeClient() 来关闭所有相关的 Socket 和字节流。

需要注意的是一个 java.lang.Thread 类的对象代表一个线程。多线程的建立就是指多个 Thread 对象在服务器开始到关闭的过程中操作,由于线程代码段在 run 方法中,当 run 方法执行完后,线程也就相应的结束了。

2. 实现线程创建所需要的 Java 技术

在传统的程序语言中,程序在运行时按照一定的流程来进行,一次只能运行一个程序块。Java 的多线程打破了这种束缚,程序可以同时运行多个程序块,使程序运行效率更高。

在项目中,当一个线程创建了,ServiceThread 类里面的 run() 方法自动启动程序并调用处理客户端的方法 handleClient()。handleClient() 需要通过与服务器接收和发送信息,按信息类型用选择语句进行登录、查询、订票等动作。在用户事件结束后通过方法 closeClient() 来关闭所有相关的 Socket 和字节流。

值得注意的是,在进行多线程和网络搭建的时候,应多使用抛异常方法 Exception 来处理代码中有可能的错误。

3. 服务器线程的搭建设计与流程

根据实现服务器线程相关 Java 技术的分析,我们可以创建一个服务器线程类 ServerAdmin 来搭建具有线程功能的服务器,类图设计如图 6-1。服务器线程处理流程如下:

（1）用户通过界面窗体发送操作命令；
（2）当该用户第一次发送命令，服务器新建线程，执行 run 方法；
（3）用户通过发送和接收信息与服务器进行交互，期间其他的用户也可以新建自己的线程（执行新的 run 方法）并进行交互；
（4）用户访问结束，其他用户继续交流直到所有用户访问结束。

绘制服务器接收用户的时序图如图 6-3 所示。

图 6-3 服务器线程的时序

4. 服务器线程的代码实现

【代码 6-1】线程服务器 ServerThread 的搭建

```
package server;
import info.Shuttle;
import info.User;
import java.io.IOException;
import java.io.InputStream;
import java.io.ObjectInputStream;
import java.io.ObjectOutputStream;
import java.io.OutputStream;
import java.net.Socket;
import java.util.Vector;
import messages.Message;
import messages.MessageExample;
import messages.MessageLoginAck;
import messages.MessageLoginReq;
```

```java
import messages.MessageShuttleQuery;
import messages.MessageShuttleQueryAck;
import messages.MessageTicketBook;
import messages.MessageTicketBookAck;
import server.database.CPYDDatabase;
public class ServiceThread extends Thread {
    private Socket socket = null;
    private ObjectOutputStream oos = null;
    public ServiceThread(Socket socket) {
        this.socket = socket;
    }
    public void run() {
        handleClient();
        closeClient();
    }
    private void handleClient() {
        OutputStream os = null;
        InputStream is = null;
        ObjectInputStream ois = null;
        try {
            //创建对象数据流
            os = socket.getOutputStream();
            is = socket.getInputStream();
            oos = new ObjectOutputStream(os);
            ois = new ObjectInputStream(is);
        } catch (IOException e) {
            System.out.println(e.getMessage());
            return;
        }
        //处理来自客户端的消息对象
        while (true) {
            //从对象输入数据流中读取数据对象
            Message msgReceived = null;
            try {
                msgReceived = (Message) ois.readObject();
            } catch (ClassNotFoundException e) {
                System.out.println(e.getMessage());
                continue;
            } catch (IOException e) { //Socket 出错,停止对次 Socket 的处理
                System.out.println(e.getMessage());
                return;
            }
```

```java
        //按消息类型处理
        switch (msgReceived.msgType) {
        case MSG_EXAMPLE:
            handleMsgExample((MessageExample) msgReceived, oos);
            break;
        case MSG_LOGIN_REQ:
            handleLoginRequest((MessageLoginReq) msgReceived, oos);
            break;
        case MSG_SHUTTLE_QUERY:
            handleShuttleQuery((MessageShuttleQuery) msgReceived, oos);
            break;
        case MSG_TICKET_BOOK_REQ:
            handleTicketBook((MessageTicketBook) msgReceived, oos);
            break;
        default:
          System.out.println("Uknown message received: "
                + msgReceived.msgType);
        }
      }
  }

  private void closeClient() {
      if (socket != null) {
        try {
              socket.shutdownOutput();
              socket.shutdownInput();
              socket.close();
          } catch (IOException e) {
              e.printStackTrace();
          }
      }
  }

  private boolean sendMessage(Message msg, ObjectOutputStream oos) {
      try {
          oos.writeObject(msg);
      } catch (IOException e) {
          e.printStackTrace();
          return false;
      }
      return true;
  }
```

```java
    private void handleMsgExample(MessageExample msgExample,
            ObjectOutputStream oos) {
        String data = msgExample.getData();
        System.out.println(data);
        msgExample.setData("Example Message Received: " + data);
        sendMessage(msgExample, oos);
    }

    private void handleLoginRequest(MessageLoginReq msgLogin,
            ObjectOutputStream oos) {
        boolean verifyPassed = false;
        User user = CPYDDatabase.userQquery(msgLogin.getAccount());

MessageLoginAck msgLoginAck = new MessageLoginAck(msgLogin.getAccount());
        //登陆成功的条件:用户账号存在,并且不是管理员,密码正确
        if (!(msgLogin.getAccount().equals("admin"))
                && user.getName().equals(msgLogin.getAccount())
                && msgLogin.verify(user.getPwd())) {
            verifyPassed = true;
            msgLoginAck.setFailReason("用户信息正确!");
        } else {
            msgLoginAck.setFailReason("用户信息不正确!");
        }
        System.out.println("Account: " + msgLogin.getAccount()
                + "Verification" + verifyPassed);

        msgLoginAck.setLoginResult(verifyPassed);
        msgLoginAck.setUser(user);

        sendMessage(msgLoginAck, oos);
    }

    private void handleShuttleQuery(MessageShuttleQuery msgShuttleQuery,
            ObjectOutputStream oos) {
      Vector<Shuttle> shuttles = CPYDDatabase.shuttleQquery(
                msgShuttleQuery.getStarting(), msgShuttleQuery.getEnding(),
                msgShuttleQuery.getYear(), msgShuttleQuery.getMonth(),
                msgShuttleQuery.getDate());

        MessageShuttleQueryAck msgAck = new MessageShuttleQueryAck();

        msgAck.setShuttles(shuttles);
```

```
        sendMessage(msgAck, oos);
    }

    private void handleTicketBook(MessageTicketBook msgTicketBook,
            ObjectOutputStream oos) {
        MessageTicketBookAck msgAck = new MessageTicketBookAck(
            msgTicketBook.getShuttleId(), msgTicketBook.getUserId());
      if (CPYDDatabase.ticketBook(msgTicketBook.getShuttleId(),
            msgTicketBook.getUserId())) {
          msgAck.setSuccess(true);
      }
sendMessage(msgAck, oos);
  }
}
```

6.2 车票预订系统的线程同步问题

在多线程的使用中,由于多个线程需要共享数据资源,如果不加以控制可能会产生冲突,这就涉及线程的同步与资源共享的问题。在项目中,多线程的车票预订系统很有可能出现车票预订人数多于车票可购买数的情况,为了避免造成数据的错误,所以我们使用 synchronized 关键字来给对象或代码块加锁,使得同一时刻最多只有一个线程执行这个段代码。

1. 线程同步实现过程

客户端发送的 Message 往往都是一直在访问共享资源,所以要实现线程同步就必须对发送命令进行同步处理,包括:

(1) 对访问共享资源的方法进行同步处理;

(2) 对访问共享资源的某一段代码进行共享处理。

2. 线程同步实现所需 Java 技术

线程的同步一般通过关键字 synchronized 给对象或类加锁,分为对象锁和类锁。这个关键字保证一段代码同时只能有一个线程在执行,保证别人不会看到对象处于不一致的状态中。对象将从一种一致的状态转变到另一种一致的状态,后来的线程将会看到后一种状态。具体来说对象锁只是加在当前语句块上,对别的语句块没有影响,这种加锁一般都是把这个关键字用在方法和同步块上面。要实现方法的线程同步,需要使用关键字 synchronized, synchronized 定义语法:

[访问修饰符] synchronized 方法类型、方法名(参数表)

在同步的过程中,我们对某一段代码进行共享处理的方法如下:

```
synchronized(Object){
        //访问 object 对象
}
```

一个方法使用 synchronized 关键字修饰后,当一个线程调用该方法时,必须先获得对象锁,只有在获得对象锁以后才能进入 synchronized 方法。一个时刻对象锁只能被一个线程持有。如果对象锁正在被一个线程持有,其他线程就不能获得该对象锁,其他线程就必须等待持有该对象锁的线程释放锁。

类锁就是把锁加在这个类上面,所有的其他对象想访问这个类的对象的任何同步方法,必须获得这个锁。这种锁一般把这个关键字用在静态方法中,一般我们很少用类锁。

如果类的方法使用了 synchronized 关键字修饰,则称该类对象是线程安全的,否则是线程不安全的。

3. 线程同步流程设计

通过 Java 相关技术分析,在项目里面,ClientDummy 类通过创建布尔值型 connectToServer() 来访问服务器 serverAdmin。如果访问成功,ClientDummy 类发送 sendMessage() 方法。在访问的过程中,由于发送和接收信息都已经加上锁,程序一次只能处理一句信息,其他信息在等待。当当前信息发送完毕,其他信息才可被处理。最后,ClientDummy 类通过方法 disconnectFromServer() 关闭套接字和所有数据流,绘制班车查询及订票的时序流程如图 6-4 所示。

图 6-4 服务器多线程的同步时序流程

4. 线程同步代码设计

在项目中,对 ClientDummy 类里面需要访问共享资源的方法进行同步处理,如【代码 6-2】。

【代码 6-2】对 ClientDummy 方法进行修改:

```
private static synchronized boolean sendMessage(Message msg){
    try{
        oos.writeObject(msg);
    }catch(IOException e){
        e.printStackTrace();
```

```
            return false;
        }
        return true;
    }
    private static synchronized Message receiveMessage() {
        try {
            return (Message) ois.readObject();
        } catch (ClassNotFoundException e) {
            e.printStackTrace();
        } catch (IOException e) {
            e.printStackTrace();
        }
        return null;
    }
```

实验 07　多线程编程

实验目的
（1）掌握 Java 线程的概念。
（2）掌握 Java 的线程搭建。
（3）掌握线程的状态与生命周期。
（4）了解线程的死锁。
（5）提高编程的逻辑思维能力和程序排错能力。
实验内容
在实验 6 的基础上，完成以下内容：学生请假管理系统的多线程的搭建。
实验安排
学生请假管理系统的多线程的搭建：3 学时。
实验报告
按照规定格式提交本次实验内容的结果。

第二部分 综合实训

通过本教材第一部分增量式项目的设计及实现过程，读者学到了增量式项目开发方法、Java 基本技能以及独立进行软件设计和开发的能力。在此基础上，本教材第二部分通过介绍多个不同类型的综合案例，进一步提高读者的编程能力与知识应用能力、团队协作能力、分析与解决新问题的能力等。

本部分共分三章，第 7 章介绍基于 C/S 结构的办公管理系统的设计与实现过程、第 8 章介绍基于 Web 的考勤系统的设计与实现过程、第 9 章介绍基于 Android 平台的应用开发。限于篇幅有限，以上三章只列出了部分核心代码，读者如有需要可联系本书作者。

第7章 办公管理系统

本章的目标是实现一个 C/S 结构的局域网办公管理系统，包括考勤签到、公告、聊天通信以及系统管理等四个主要功能。本系统在局域网内部署后，可供多个用户同时使用。在系统实现中，涉及到多台计算机通过服务器进行通信的问题，此类通信要保证可靠性和安全性，一般都是基于 TCP/IP 协议实现。

本系统涉及到的 Java 技能包含 GUI 编程与事件处理、集合类、TCP Socket 套接字通信、对象序列化、对象流、多线程和数据库编程等。

7.1 需求分析与项目目标

办公管理系统的用户主要是公司内部员工，包括高级职员（如经理）和普通员工，其功能一般包括考勤签到、公告、聊天通信以及系统管理等四个模块。本章需要实现如图 7-1 所示的办公管理系统客户端主界面。

（1）考勤模块，主要功能包括：员工签到和签到历史信息查询。

（2）公告模块，主要功能包括：查看公告、删除公告和发布公告。

（3）通信模块，主要功能包括：单人通信、上下线状态显示、离线信息发送、多人通信。

（4）系统管理模块，主要功能包括：查看个人信息、修改个人信息、注册员工、删除员工、查看员工信息。

图 7-1 办公管理系统主界面

7.1.1 用例分析

用户使用办公管理系统一般要经历以下几个步骤：首先启动办公管理系统，然后在成功登录后，根据自己拥有的权限可以在考勤模块、通信模块、公告模块、管理模块中使用相应的功能，最后可在使用完毕后关闭程序界面。图 7-2 和 7-3 分别列出了不同角色（普通员工和经理）所拥有的权限和能使用的功能，表 7-1 给出了各个用例的详细描述。

图 7-2 普通员工权限用例分析

图 7-3 经理权限用例分析

表7-1 局域网管理系统用例描述

用例	用例描述	
	用户操作	软件功能
用例1	启动服务器	显示服务器主界面； 界面上显示【开启服务器】和【关闭服务器】按钮以及显示需要使用的端口号
用例2	打开客户端并登录	显示客户端主界面； 在界面上输入账号、密码后登录，登录时的前提条件是连接服务器成功，登录时需要通过服务器端并连接相应的数据库判断用户名和密码正确与否，正确则进入功能界面并且得到相应的权限，否则弹出用户或密码错误信息
用例3	考勤签到	正常签到则弹出签到成功，否则弹出今天已签到
用例4	查看考勤	根据相应的ID账号和时间可以查看到不同员工每月的签到天数，若单击【清空】按钮可清空全部已经签到过的记录
用例5	查看及删除公告	进入查看界面后有【删除公告】、【上一页】和【下一页】按钮； 能够查看到全部已经发布的公告及发布的时间； 单击相应公告的标题则可以查看到详细的公告内容； 选中标题前面的正方形框，单击【删除】按钮则可以删除选中的公告； 若公告条数超过一页可按【下一页】查看，如果已经是最后一页则显示"已经是最后一页"信息
用例6	发布公告	在界面中的标题框和正文框内输入相应的信息后单击【发布公告】按钮，则会把数据发送到服务器端并存入数据库
用例7	聊天上下线	若一个好友上线了，则其他相应好友的客户端头像将会变亮，好友下线头像会变暗
用例8	聊天发送消息	双击【好友姓名】后，在输入框内输入相应的内容单击【发送】按钮即可
用例9	聊天信息接收	若发送者发送信息给对方时，如果接收方不在线，则会在接收方上线时，发送者姓名变为红色以便提示接收方有留言。若接收方在线则发送者头像直接变红
用例10	多人通信	添加群组：在弹出的输入框中输入群组名字即可进群； 删除群组：单击相应的群组名使名字变蓝，然后单击【删除群组】按钮即可； 群组通信：将鼠标移动到你要发送的群组名上，名字会变粉红。双击则会弹出群组消息框，在相应消息框中输入要发送的信息后单击【发送】按钮即可
用例11	查看个人信息	在主界面中单击【查看个人信息】按钮则弹出个人信息

续表 7-1

用例	用例描述	
	用户操作	软件功能
用例 12	修改个人信息	在主界面中单击【修改个人信息】按钮,然后再单击【修改】按钮,在相应输入框内输入要修改的信息,然后单击【确认修改】按钮即可
用例 13	注册员工	在主界面中单击【注册员工】按钮并在相应的输入框中输入要注册的信息后单击【确定】按钮即可
用例 14	删除员工	进入相应界面,单击要删除的员工名单,名字就会显示在右边,然后单击【删除】按钮即可
用例 15	删除员工实现分页	在删除员工界面通过单击【上一页】和【下一页】按钮实现
用例 16	查看员工信息	直接查看:界面左下角会显示相应的员工名单,单击要查看信息的员工名字即可查看到员工信息; 搜索查看:在输入框中输入要查询的名字,每输入一个字,左下角的员工名单都会作出相应的改变,可以选择要搜索的名字即可显示出相应的信息

7.1.2 需求分析

根据表 7-1 的用例描述,表 7-2 列出了具体的需求条款及需求描述。

表 7-2 办公管理系统需求列表

需求编号	需求描述	解释
Req7-1	服务器端界面上有【开启服务器】和【关闭服务器】两个按钮	用例 1
Req7-2	客户端登录界面有【账号】、【密码】输入框和【登录】按钮,并且客户端登录界面有背景图片	用例 2
Req7-3	验证账号及登录密码的正确性	用例 2
Req7-4	密码错误可以弹出【密码错误】提示框	用例 2
Req7-5	密码正确则显示功能选择界面,并根据相应权限保证不同功能的权限访问	用例 2
Req7-6	功能界面有各个功能的 GUI 按钮及对应的背景图片	用例 2
Req7-7	签到后能够提示签到成功或者今天已签到	用例 3
Req7-8	实现考勤界面的 GUI	用例 4
Req7-9	提供两种查看考勤记录的方法:账号和年份	用例 4
Req7-10	清除考勤记录	用例 4
Req7-11	实现查看公告功能的 GUI	用例 5
Req7-12	显示公告及公告发布时间	用例 5

续表 7-2

需求编号	需求描述	解释
Req7-13	删除公告	用例 5
Req7-14	查看上一页及下一页公告	用例 5
Req7-15	实现发布功能的 GUI	用例 6
Req7-16	发布公告	用例 6
Req7-17	实现聊天功能的 GUI	用例 7
Req7-18	显示好友上下线头像的变化	用例 7
Req7-19	实现单人聊天通信	用例 7
Req7-20	发送留言信息	用例 7
Req7-21	发送在线消息	用例 7
Req7-22	提示收到消息使好友的头像变红	用例 7
Req7-23	实现多人聊天	用例 7
Req7-24	实现管理功能的全部 GUI	用例 11～16
Req7-25	查看个人信息	用例 11
Req7-26	修改个人信息	用例 12
Req7-27	注册新员工	用例 13
Req7-28	删除员工	用例 14
Req7-29	删除员工界面具有分页效果	用例 15
Req7-30	经理直接查看员工信息	用例 16
Req7-31	经理通过搜索、查找、查看员工信息	用例 16

7.1.3 项目目标

本章的设计目标是设计一个局域网办公管理系统，其主界面效果如图 7-1 所示，由于通信功能在《Java 程序设计及应用——增量式项目驱动一体化教程》中的局域网聊天工具中进行了详细介绍，所以通信功能的实现不作为本次项目目标。因此，本章中办公管理系统的项目目标包含：实现图 7-2 和图 7-3 中除去聊天功能之外的所有用例，满足表 7-3 中包含的 Req7-1 至 Req7-16、Req7-24 至 Req7-31 的需求条款。

7.2 功能分析与软件设计

本节在需求分析的基础上，首先从功能的角度分析软件的实现流程和软件设计过程，然后设计用户数据库表格以及相应的程序接口。

7.2.1 界面效果设计

本系统的界面元素主要通过背景图片和一些处理过的按钮图片组成，背景图片经过PhotoShop处理后可以达到更好的显示效果，如图7-4所示【登录】按钮和【查看个人信息】按钮是通过photoshop处理后实现的界面效果。

图7-4 【登录】、【查看个人信息】按钮

首先准备三张经过处理的图片素材（如图7-5所示），然后编写Java代码以在界面上显示具有背景图片的按钮，具体实现代码见【代码7-1】。

图7-5 【登录】图片素材

【代码7-1】在界面上显示背景图片

```
JButton enter = new JButton();
//鼠标经过的时候的图片
ImageIcon entered = new ImageIcon("src/image/button1.png");
//初始图片
ImageIcon released = new ImageIcon("src/image/button2.png");
//按下去的图片
ImageIcon pressed = new ImageIcon("src/image/button3.png");
enter.setOpaque(false); //透明化
enter.setIcon(released); //初始化
enter.setBorderPainted(false);
enter.setBorder(null);
enter.setBounds(267,290,90,32); //按钮大小,位置
//按钮透明,这样用圆形按钮的时候也不会出现按钮边框
enter.setContentAreaFilled(false);
//按下按钮时的图片
enter.setPressedIcon(pressed);
//经过时的图片
enter.setRolloverIcon(entered);
```

7.2.2 登录功能

用户进入客户端主界面时需要登录（如图7-6所示），登录过程如图7-7所示。在客户端主界面，用户填写完账号和密码并单击【登录】按钮后，客户端将通过TCP Socket与服务器端进行连接；服务器通过读取数据库中保存的账号和密码来验证登录的合法性，如果验证通过，服务器则在数据库中记录相应的Socket信息，否则不做记录；

服务器把登录验证的结果返回给客户端,如果验证通过,客户端进入功能主界面,否则显示密码错误等提示信息。

图7-6 登录界面

图7-7 登录过程

7.2.3 管理模块

用户成功登录后,就可以通过管理模块管理相关信息。对于普通员工来说,只具有查看个人信息的功能;对于经理用户来说,包含查看个人信息、修改个人信息、注册员工、删除员工和查看员工信息等功能。

1. 查看个人信息

用户登录成功后，可以单击【管理】功能模块的【查看个人信息】按钮查看用户自己的信息。

查看个人信息的流程如图7-8所示，当用户单击【查看个人信息】按钮后，客户端（如图7-1所示）会向服务器发出查询请求；服务器根据用户的ID号从数据库中读出用户信息，并发送回客户端；最后客户端把收到的信息显示在如图7-9所示的个人信息界面中。

图7-8 查看个人信息流程

图7-9 查看个人信息界面

2. 修改个人信息

用户成功登录后，可以单击【管理】功能模块的【修改个人信息】按钮修改用户自己的信息。

修改个人信息的流程如图7-10所示。首先，用户单击【修改个人信息】按钮；然后客户端根据不同的用户权限显示不同的界面效果，若是普通员工，则禁止修改，若用户是经理，则可以做相应的修改。进入如图7-11所示的个人信息修改界面后，客户端需要从服务器端的数据库内读取相应的用户数据,并显示在用户信息界面中,在该界面中用户只需要单击【修改】按钮并在相应的文本框内输入要修改的信息,单击【确认修改】按钮即可;这时客户端会把更新的个人信息发送给服务器,服务器负责把新的个人信息更新到数据库。

图 7-10　修改个人信息流程

图 7-11　修改个人信息界面

3. 注册员工

经理用户成功登录后，可以单击【管理】功能模块的【注册员工】按钮注册新的员工。

注册员工的流程如图7-12所示。首先用户单击【注册员工】按钮，然后客户端判断用户权限，若是普通员工则不做响应，若是经理用户则进入【注册员工】界面（如图7-13所示）。填写完相关注册信息后，单击【确定】按钮，客户端会通过Socket连接，采用序列化技术把员工信息发送到服务器端，服务器负责把新的员工信息写入数据库。

图7-12 注册员工流程

图7-13 注册员工界面

4. 删除员工

经理用户成功登录后，可以单击【管理】功能模块的【删除员工】按钮删除员工信息。

删除员工的流程如图 7 – 14 所示。当用户单击【删除员工】按钮时，客户端要判断用户权限，若是普通员工则不做响应，若是经理用户则显示删除员工界面（如图 7 – 15 所示）。在删除界面，用户选中待删除的员工记录后，单击【确定】按钮，客户端会通过 Socket 连接和序列化技术把员工信息发送到服务器端，服务器负责把该员工的记录从数据库中删除。

图 7 – 14　删除员工流程

图 7 – 15　删除员工界面

5. 查看员工信息

经理用户登录成功后，可以单击【管理】功能模块的【查看员工信息】按钮查看员工信息。

查看员工信息的流程如图 7-16 所示。用户单击【查看员工信息】按钮后，客户端判断用户权限，若是普通员工则不做响应，若是经理则显示【查看员工信息】界面（如图 7-17 所示），并在界面上显示员工列表，经理可以通过搜索的方式或通过选择用户记录的方式直接查看员工信息。如果员工比较多，用户可以在输入框内填写待查看的员工姓名，客户端会自动筛选用户名记录；如果只有一条记录，经理可以单击【查看】按钮，客户端就会通过 Socket 连接到服务器并发送一个消息，请求服务器返回对应的员工信息，服务器从数据库中搜索具体员工的信息后并返回给客户端；最后客户端再从相应的消息中读出数据显示在【查看员工信息】界面的右半部分，效果如图 7-18 所示。经理也可以从员工列表中找到员工记录，双击即可查看员工信息。

图 7-16　查看员工信息流程

7.2.4　考勤模块

考勤模块需要实现客户端主界面（如图 7-1 所示）中的"考勤"功能，包括"签到"和"查看考勤"两个子功能。

1. 签到

用户签到的流程如图 7-19 所示。用户登录成功后，单击【签到】按钮进行签到，客户端会向服务器端的数据库验证查看是否已经签到，如果员工是当天第一次签到，客

图7-17 查看员工信息界面

图7-18 查看员工详细信息示例

户端会发送给服务器一个签到消息,并且显示【签到成功】对话框(如图7-20所示),服务器会在数据库中更新此员工的签到信息;如果该员工当天已经签到了,客户端则会弹出【你今天已经签到】对话框(如图7-21所示),提示用户今天已签到。

图 7-19 用户签到流程

图 7-20 签到成功对话框

图 7-21 已签到对话框

2. 查看考勤

用户查看考勤的流程如图 7-22 所示。用户单击【查看考勤】按钮后，客户端会判断用户权限，若为普通员工则提示权限不足，若为经理则进入【查看考勤】界面。用户可以通过选择相应的账号和时间查看考勤信息。当用户选择完成后，客户端会向服务器端发送查看考勤记录请求，服务器从数据库中读取指定信息后发送给客户端，并在客户端的界面上显示考勤记录。当用单击【清空记录】按钮时，客户端会将指定的员工账号和时间发送给服务器，服务器收到消息后将更新数据库。

图 7-22 查看考勤和清空考勤流程

图 7-23 查看考勤界面

111

7.2.5 公告模块

用户登录成功后，通过单击【查看公告】按钮查看公司的内部公告。对于经理用户来说，还可以【发布公告】和【删除公告】。

1. 查看公告

软件的查看公告和删除公告流程如图 7-24 所示。在客户端，用户单击【查看公告】按钮查看公告，客户端首先向服务器端发送查看公告请求消息；服务器从数据库中读取公告标题信息，并返回给客户端；在客户端界面上显示公告列表（如图 7-25 所示），用户单击任一公告标题即可查看公告内容（如图 7-26 所示）。

图 7-24　公告查看、删除公告流程

图 7-25　查看公告

图 7 - 26 查看公告内容

在公告列表中，用户可以选择一条或者多条公告，单击【删除】按钮可删除公告。由于只有经理权限才能删除公告，客户端首先会验证用户的权限，如果为普通员工，则不予删除；如果为经理权限，客户端会把指定公告条款的信息发送给服务器，服务器收到删除信息的请求后，从数据库中删除相应记录即可。

2. 发布公告

发布公告的流程如图 7 - 27 所示。用户单击【发布公告】按钮，由于只有经理权限才能发布公告，所以客户端首先会检查用户权限，如果权限为普通员工，发布公告操作将没有反映，如果权限为经理级别，客户端会弹出发布公告界面（如图 7 - 28 所示）。在用户填好公告单击【发布公告】按钮时，客户端会把新创建公告的信息发送给服务器，服务器收到请求后，把公告信息更新进数据库。

7.2.6 通信模块

本节仅对通信模块需要实现的功能进行分析，在增量开发阶段可不予实现。

1. 上下线通知及界面初始化

当用户单击【通信】模块中的【聊天】按钮时，客户端会弹出如图 7 - 29 所示的聊天界面。在该界面的【好友列表】和【群组列表】中，如果用户名显示为黑色字体，则表示用户处于在线状态；如果用户名显示为灰色字体，则表示用户处于离线状态。如果好友的状态发生变化，【好友列表】和【群组列表】中也会及时更新。

好友上下线的软件流程如图 7 - 30 所示。具体的通信过程为：在客户端弹出如图 7 - 29 所示的聊天主界面后，客户端向服务器通过 Socket 发送一系列的用户请求，依次

图7-27 发布公告流程

图7-28 发布公告界面

为:好友列表信息、群组列表信息和离线好友消息;服务器收到请求后,根据请求的类型,到数据库中查询,并把查询结果依次通过Socket返回给客户端;客户端把收到的好友信息、群组信息和离线好友消息一次性更新到聊天界面中。

图7-29 聊天主界面

图7-30 好友上下线的软件流程

如果用户状态发生变化，服务器则通过检测到相应 Socket 的开关状态，将用户在数据库中的信息和状态进行更新，并把相应用户及其状态的变化发送给好友和群组。当客户端收到好友信息变化的请求时，就可以在好友列表和群组列表中进行更新。当用户下线时，客户端会发送下线消息给服务器，服务器更新数据库中对应用户的信息，并把下线消息发送给相应在线好友客户端。如果用户处于离线状态，服务器会把相应的离线个人信息和群组聊天信息保存进数据库，等用户上线时再发送给客户端。

2. 单人通信

当用户双击好友（这里也称为头像），客户端会弹出如图 7 – 31 所示的单人通信聊天框，若双击好友时好友名为红色，聊天框中会显示未读信息。

图 7 – 31　单人通信聊天框

好友上下线流程如图 7 – 32 所示。首先，用户在输入框内输入想要发送的消息，然后单击【发送消息】按钮进行消息的发送。客户端在发送消息前会判断好友是否在线，如果好友在线则发送聊天消息给好友的客户端；如果好友离线，则把聊天消息发送给服务器，服务器会把离线消息保存进数据库，等该好友上线时再发给该好友的客户端。如果好友在下线前已经读取了聊天消息，则这些消息不会存入数据库；如果好友在下线前未能读取消息，服务器则会把这些消息存进数据库待好友下次上线时接收。从技术实现上，本部分用到了对象序列化技术，实现消息对象在 Client 端与 Server 端进行聊天消息的通信。

3. 多人通信

多人通信的原理与单人通信基本相同，感兴趣的读者可以参考实现。

7.2.7　数据库表格与脚本设计

局域网办公管理系统数据库（取名 office）的设计需要以用户为中心，数据库表包含员工信息表（以员工 ID 为主键）、用户聊天消息记录表（以员工 ID 为主键）、用户群组表、公告表（以公告 ID 为主键）和员工签到表（以签到记录 ID 为主键，以员工 ID 为外键），所有表格汇总于表 7 – 3。

图 7-32　单人通信设计流程

表7-3 局域网办公管理系统数据库表汇总

表名	功能说明
员工信息表	用于查询、更新员工信息
聊天消息记录表	用来存放讨论消息
群组表	用于存放群组
公告表	用于存放公告
签到表	用于签到

表7-4给出了员工信息表的字段描述，代码7-2给出了创建该表的数据库T-SQL脚本。

表7-4 员工信息表 userinfo

字段名称	属性名	类别	主键	非空
用户ID	id	VARCHAR（16）	是	是
用户权限	authority	VARCHAR（10）	否	是
用户密码	pass	VARCHAR（16）	否	是
薪水	income	INT	否	是
年龄	age	INT	否	是
性别	sex	VARCHAR（4）	否	是
地址	address	VARCHAR（100）	否	是
电话	telpNum	VARCHAR（15）	否	是
QQ号码	QQNum	VARCHAR（15）	否	是

【代码7-2】创建 userinfo

```
create table userinfo(
    id VARCHAR(16) primary key,
    authority VARCHAR(10) NOT NULL,
    pass VARCHAR(16) NOT NULL,
    income INT NOT NULL,
    age INT NOT NULL,
    sex VARCHAR(4) NOT NULL,
    address VARCHAR(100) NOT NULL,
    telpNum VARCHAR(15) NOT NULL,
    QQNum VARCHAR(15) NOT NULL
);
```

表7-5给出了聊天消息记录表的字段描述，代码7-3给出了创建该表的数据库T-SQL脚本。

表 7-5 聊天消息记录表 message

字段名称	字段名称	类别	主键	非空
发送 ID	SenderID	VARCHAR（16）	否	是
接收者 ID	GetterID	VARCHAR（16）	否	是
信息内容	Cont	VARCHAR（150）	否	是

【代码 7-3】创建 message

```
create table message(
    SenderID VARCHAR(16) primary key NOT NULL,
    GetterID VARCHAR(16) NOT NULL,
    Cont VARCHAR(150) NOT NULL
);
```

表 7-6 给出了群组表的字段描述，【代码 7-4】给出了创建该表的数据库 T-SQL 脚本。

表 7-6 群组表 groups

字段名称	字段名称	类别	主键	非空
用户 ID	id	VARCHAR（16）	否	是
组名	groupName	VARCHAR（40）	否	是

【代码 7-4】创建 groups

```
create table groups(
    id VARCHAR(16) primary key not null,
    groupName VARCHAR(40) not null
);
```

表 7-7 给出了群组表的字段描述，【代码 7-5】给出了创建该表的数据库 T-SQL 脚本。

表 7-7 publication

字段名称	字段名称	类别	主键	非空
标题	title	VARCHAR（40）	否	是
信息内容	message	TEXT	否	是
发布时间	time	timestamp	否	是

【代码 7-5】创建 publication

```
create table publication(
    title VARCHAR(40) primary key,
    message TEXT not null,
    time timestamp default current_timestamp on update current_timestamp
);
```

表7-8给出了签到表的字段描述，【代码7-6】给出了创建该表的数据库TSQL脚本。

表7-8 签到表 attend

字段名称	字段名称	类别	主键	非空
签到ID	id	INT	是	是
用户ID	userid	VARCHAR（16）	否	是
签到时间	time	timestamp	否	是
是否签到	present	VARCHAR（10）	否	是

【代码7-6】创建 attend

```
create table attend(
    id INT primary key not null,
    userid VARCHAR(16) not null,
    time timestamp default current_timestamp on update current_timestamp,
    present VARCHAR(10) null
);
```

7.2.8 增量开发计划

通过上述功能分析，我们已经熟悉每个功能的实现流程，本部分将制订一个开发计划，按照增量开发的方式实现该系统。增量开发计划以功能界面为出发点，通过功能的迭代逐步完成客户端和服务器端。

表7-9 增量开发计划

增量#	功能实现	对应用例	需求列表	相关技术
增量7-1：搭建系统主体架构	1. 服务器端界面； 2. 客户端登录与功能选择界面； 3. 服务器与客户端Socket连接； 4. 实现登录功能，能够验证密码正确性	用例1 用例2	Req7-1 至7-6	GUI编程 事件处理 多线程技术 序列化技术 对象流技术 Socket通信
增量7-2：实现员工管理功能	1. 查看个人信息 2. 修改个人信息 3. 注册员工 4. 删除员工 5. 查看员工信息	用例11 用例12 用例13 用例14 用例15 用例16	Req7-25 至7-31	GUI编程 事件处理 Socket通信 消息包Message 对象序列化 对象流操作 多线程

续表 7-9

增量#	功能实现	对应用例	需求列表	相关技术
增量 7-3：实现公告	实现公告	用例 5 用例 6	Req7-11 至 7-16	GUI 编程 事件处理 多线程技术 序列化技术 对象流技术 Socket 通信
增量 7-4：实现考勤功能	实现考勤	用例 3 用例 4	Req7-7 至 7-10	GUI 编程 事件处理 多线程技术 序列化技术 对象流技术 Socket 通信

7.3 项目增量开发

本节按照增量开发计划依次实现各个增量，按照增量的实现过程最终实现局域网办公管理系统的预定功能。

7.3.1 增量 7-1：搭建系统主体架构

增量 7-1 实现的是用例 1 和用例 2，实现的功能包括服务器端界面、客户端登录与功能选择界面、服务器与客户端 Socket 连接和客户登录验证等功能，为后续增量提供一个端到端（客户端到服务器端）的开发和测试环境。

参照如图 7-7 所示的用户登录流程，增量 7-1 实现了客户端和服务器端的主界面，以及 Socket 的通信过程，用户身份验证序列见图 7-33，具体包括以下几个步骤：

（1）服务器端界面和事件处理；
（2）服务器端 Socket 处理；
（3）数据库和数据库编程接口；
（4）客户端服务线程；
（5）客户端登录界面、登录过程和功能选择界面。

1. 服务器端：实现服务器端界面

本节需要实现如图 7-34 所示的服务器界面，左图是服务器尚未开启的界面效果，右图是用户单击【开启服务器】按钮后的界面效果，开启服务器后则进入服务状态，客户端可以发送 Socket 连接请求。

我们定义 ServerUI 类（代码 7-7）实现图 7-34 的界面效果和事件处理，定义 StartServer 类（代码 7-8）作为服务器端主程序的入口。ServerUI 的 run（）方法创建了服务器的 ServerSocket 监听进程 ServerMainThread（参见代码 7-9），此进程负责创建与

Java 程序设计实训——增量式项目驱动一体化教程

图 7-33 用户验证序列

客户端的 Socket 连接和用户验证。

图 7-34 服务器界面：左图为未开启服务进程界面，右图为已开启服务进程界面

【代码 7-7】服务器界面 ServerUI 类

```
package server;
import java.awt.event.*;
import javax.swing.*;
/* 服务器的 GUI */
public class ServerUI extends JFrame implements ActionListener, Runnable {
    private JButton JB_Open, JB_Close;
```

```java
    public ServerUI() {
        GUI();
    }
    public void GUI() {
        ImageIcon image = new ImageIcon("src/服务器背景图.jpg");
        JLabel jl = new JLabel(image);
        jl.setBounds(0,0,200,320); this.add(jl);
        this.setTitle("服务器");
        this.setSize(200,320);this.setLocation(600,100);
        this.setLayout(null);
        this.setDefaultCloseOperation(EXIT_ON_CLOSE);
        JB_Open = new JButton("开启服务器");
        JB_Close = new JButton("关闭服务器");
        JB_Open.setBounds(45,80,100,45);
        JB_Close.setBounds(45,180,100,45);
        jl.add(JB_Open); jl.add(JB_Close);
        JB_Open.addActionListener(this);
        JB_Close.addActionListener(this);
        this.setVisible(true);
    }
    public void actionPerformed(ActionEvent e) {
        if (e.getSource() == JB_Open) {
            new Thread(ServerUI.this).start();
            JB_Open.setEnabled(false);
        }
        if (e.getSource() == JB_Close) {
            System.exit(0);
        }
    }
    public void run() {
        int port = Integer.valueOf(1);
        new ServerConnClient(port);
    }
}
```

【代码7-8】服务器端应用程序 StartServer 类（用于启动服务器）

```java
package server;
public class StartServer {        /*开始运行*/
    public static void main(String[] agrs) {
        new ServerUI();
    }
}
```

2. 服务器端：实现服务器端的 Socket 通信和密码验证

为了实现能够同时为多个客户端提供服务的服务器端，需要采用多线程技术，分别实现服务器主线程和多个客户端服务线程。在服务器主线程与客户端建立连接后，服务器主线程为每一个客户端实例建立一个服务线程，由该服务线程处理客户端的数据请求，包括消息处理、消息转发、数据库访问等。

定义 ServerMainThread 类（代码 7-9），通过 TCP 网络编程技术，开启服务器端的 Socket，监听来自客户端的连接请求，并为每一个客户端启动一个单独的服务线程 ClientHandler（代码 7-12），此线程启动之后，服务器与客户端之间的通信均由其处理。

ServerMainThread 类首先对客户端发送过来的 Socket 进行连接；然后使用对象数据流技术读出 User 对象信息 [注：User 类（代码 7-10）是序列化类，在客户端和服务器端均包含该类的实现]；再把 User 对象信息发送给数据库判断账号和密码是否正确，并通过一个 result 变量返回验证结果，如果账号和密码正确，则从数据库中读取相应权限，通过 User 类记录用户验证结果和权限；最后采用对象数据流返回一个 User 对象给客户端。客户端的验证通过 JDBC 类（代码 7-11）来完成，需要预先在服务器端配置数据库 office（使用 7.2.7 节中的 SQL 脚本来创建）。

如果用户验证通过，服务器会启动一个 ClientHandler 线程处理后续的对象数据流通信。同时，服务器把该客户端服务线程加入客户服务线程管理器 ClientHandlerManager 中（代码 7-14），由客户服务线程管理器负责管理客户端的 Socket 信息，这样服务器就可以通过该管理器来简化对用户的管理。

客户服务线程 ClientHandler，首先从 Socket 对象中读取对象输入流，然后每次从该对象输入流中读取 Message（代码 7-13）对象，再针对不同消息进行不同处理。

【代码 7-9】ServerMainThread 类

```java
package server;
import java.io.*;
import java.net.*;
import clienthandlers.ClientHandler;
import clienthandlers.ClientHandlerManager;
import common.*;
/* 服务器开启了与 client 端连接的端口 */
public class ServerMainThread {
    public ServerMainThread(int port) {
        try {
            ServerSocket ss = new ServerSocket(port);
            while (true) {
                Socket s = ss.accept();
                //首先验证用户的合法性
                ObjectInputStream ois = new ObjectInputStream(s.getInputStream());
                User user = (User) ois.readObject();
                boolean result = new JDBC(user.getId(), user.getPass()).judge();
```

```java
            ObjectOutputStream oos = new ObjectOutputStream(s.getOutputStream());
                User reUser = new User();
                if(result == true){
                    reUser.setType(true);
                    String authority = new JDBC().judgeAuthority(user.getId());
                    reUser.setAuthority(authority);
                    oos.writeObject(reUser);
                    //把SOCKET添加管理类
                    ClientHandler clientHandler = new ClientHandler(s);
                       ClientHandlerManager.addClientHandler(user.getId(), clientHandler);
                    Thread thread = new Thread(clientHandler);
                    thread.start();
                }
                else{
                    reUser.setType(false);
                    oos.writeObject(reUser);
                }
            }
        }catch(Exception e){
            e.printStackTrace();
        }
    }
}
```

【代码7-10】User 类

User 类是一个实现了对象序列化的用户信息存储类, 其中 id 用于存储账号, pass 用于存储密码, authority 用于存储权限, Login_type 用于存储客户验证是否通过。

```java
package common;
import java.io.*;
public class User implements java.io.Serializable{
    private String id;
    private String pass;
    private boolean Login_type;
    private String authority;
    public boolean getType(){
        return Login_type;
    }
    public void setType(boolean Login_type){
        this.Login_type = Login_type;
    }
    public String getId(){
```

```java
        return id;
    }
    public void setId(String id){
        this.id = id;
    }
    public String getPass(){
        return pass;
    }
    public void setPass(String pass){
        this.pass = pass;
    }
    public void setAuthority(String authority){
        this.authority = authority;
    }
    public String getAuthority(){
        return authority;
    }
}
```

3. 服务器端：实现数据库 JDBC 的用户验证功能

代码 7-11 中的 JDBC 类，通过访问数据库中的用户信息，实现了用户验证和用户权限获取两个功能。其中，judge() 方法用于用户登录时的身份验证，如果验证通过，则返回用户权限和登录结果，否则返回验证失败；judgeAuthority() 方法用于在数据库中查询用户权限。

【代码 7-11】数据库接口 JDBC 类

```java
package server;
import java.sql.Connection;
import java.sql.DriverManager;
import java.sql.PreparedStatement;
import java.sql.ResultSet;
import java.sql.SQLException;
import java.sql.Statement;
import java.util.ArrayList;
import java.util.Date;
public class JDBC{
    private String driver = "com.mysql.jdbc.Driver";
    private String url = "jdbc:mysql://localhost:3306/office";
    private String userName = "root";
    private String sqlpassword = "root";
    private Connection conn = null;
    private Statement stmt = null;
    private ResultSet rs;
```

```java
    private String id;
    private String pass;
    public JDBC() { }
    public JDBC(String id , String pass){
        this.id = id;
        this.pass = pass;
    }
    //登录验证
    public boolean judge(){
        boolean result = false;
    try {
        Class.forName(driver);
        conn = DriverManager.getConnection(url, userName, sqlpassword);
        stmt = conn.createStatement();
        String sqlSelect = "SELECT id ,pass" + "FROM userInfo";
        rs = stmt.executeQuery(sqlSelect);
        while (rs.next()) {
            String DBid = rs.getString(1);
            String DBpass = rs.getString(2);
            if (DBid.equals(id) && DBpass.equals(pass)) {
                result = true;
            }
        }
        rs.close();
        stmt.close();
        conn.close();
      }catch (Exception e) {
         e.printStackTrace();
      }
        return result;
    }
    public String judgeAuthority(String id) { //判断权限
        String authority = null;
        try {
          Class.forName(driver);
          conn = DriverManager.getConnection(url, userName, sqlpassword);
          stmt = conn.createStatement();
          String sqlSelect = "SELECT id,authority" + "FROM userInfo where id = '" + id + "'";
          rs = stmt.executeQuery(sqlSelect);
          while (rs.next()) {
```

```
                authority = rs.getString(2);
            }
            rs.close();
            stmt.close();
            conn.close();
        } catch (Exception e) {
            e.printStackTrace();
        }
        return authority;
    }
}
```

4. 服务器端：建立客户端服务线程 ClientHandler

ClientHandle 类（代码 7 – 12）实现了服务器如何处理客户端发出的消息，主要负责与客户端的通信。

首先接收来自客户端的消息 Message（代码 7 – 13），如果有消息到达，则做出相应处理，如聊天信息的转发、数据库访问结果反馈等。这些消息处理功能是增量 7 – 1、增量 7 – 2 和增量 7 – 3 所要实现的。

用户验证成功后，客户端与服务器端的 Socket 以 Message 对象作为数据收发的基本单元，收发数据通过对象数据流技术实现。

由于服务器同时为多个客户端提供服务，为了便于管理客户线程 ClientHandler，服务器端提供了客户服务线程管理器 ClientHandlerManager 类（见代码 7 – 14）。ClientHandlerManager 类是一个静态类，在服务器端程序运行过程中只存在一份对象实例，该类通过一个 HashMap 容器对象来保存客户服务线程 ClientHandler 对象，在服务器主动与客户端通信的时候，服务器可以通过客户服务线程管理器，根据用户 ID 找到客户服务线程，进而取出对应的 Socket 完成通信过程。ClientHandlerManager 类的 addClientHandler () 方法在用户验证成功时使用，用于把一个客户服务线程加入 HashMap 容器对象，getClientHandler () 方法在服务器查找某个客户服务线程时使用。通过 ClientHandlerManager 类，服务器端程序可以很方便地访问任意一个客户端服务线程。

【代码 7 – 12】ClientHandler 类

```
package clienthandlers;
import java.io.*;
import java.net.*;
import common.Message;
/*这个类是专门用来接收客户端发来的消息*/
public class ClientHandler implements Runnable {
    private Socket s;
    public Socket getS() {
        return s;
    }
    public ClientHandler(Socket s) {
```

```
            this.s = s;
        }
    public void run() {
        while (true) {
            try {
            ObjectInputStream ois = new ObjectInputStream(s.getInputStream());
            Message ms = (Message) ois.readObject();
            //后续与客户端通信的代码加在这里
        }catch (Exception e) {
            e.printStackTrace();
            break;
            }
        }
    }
}
```

【代码 7-13】 Message 类

```
package common;
import java.util.ArrayList;
import java.util.Date;
public class Message implements java.io.Serializable {
    private String Content;
    private int Messtype;
    private String Sender, Getter;
    private ArrayList List;
    private Date Time;
    private String Date[];
    private int Sum[];
    public void setTime(Date time) {
            Time = time;
    }
    public Date getTime() {
            return Time;
    }
    public void setDate(String string[]) {
            Date = string;
    }
    public String[] getDate() {
            return Date;
    }
    public int[] getDay() {
            return Sum;
    }
```

```java
    public void setDay(int sum[]) {
        Sum = sum;
    }
    public ArrayList getList() {
        return List;
    }
    public void setList(ArrayList list) {
        List = list;
    }
    public String getContent() {
        return Content;
    }
    public void setContent(String content) {
        Content = content;
    }
    public int getMesstype() {
        return Messtype;
    }
    public void setMesstype(int messtype) {
        Messtype = messtype;
    }
    public String getSender() {
        return Sender;
        }
    public void setSender(String sender) {
        Sender = sender;
    }
    public String getGetter() {
        return Getter;
    }
public void setGetter(String getter) {
        Getter = getter;
    }
}
```

【代码 7-14】 ClientHandlerManager 类

```java
package clienthandlers;
import java.util.HashMap;
public class ClientHandlerManager {
    private static HashMap<String, ClientHandler> handlers = new HashMap<String, ClientHandler>();

    public static void addClientHandler(String SenderId, ClientHandler han-
```

```
dler){
        handlers.put(SenderId, handler);
    }
    public static ClientHandler getClientHandlern(String getter){
        return (ClientHandler)handlers.get(getter);
    }
}
```

5. 客户端：客户端和客户端登录过程

根据图 7-7 所示的登录流程来实现客户端的登录过程。客户端程序的入口是 StartClient 类（代码 7-15），用于启动客户端的登录界面 ClientUI（代码 7-16）。ClientUI 类实现了客户端登录界面（如图 7-6 所示），完成了客户端与服务器的 Socket 连接、客户认证（ClientAuthenticate 类，由代码 7-17 实现）等过程。如果客户认证通过，客户端会弹出办公管理系统的客户端界面 ClientMainUI（代码 7-18）。

ClientAuthenticate 类实现了客户端登录过程的身份验证。首先，创建客户端的 Socket 并连接到服务器；然后使用对象流向服务器发送一个 User 类对象；再读取一个服务器端返回的 User 对象。User 类的代码和服务器端的相同（参见代码 7-10），在服务器和客户端各有一份，并且都放在 common 包内。

从服务器接收到的 User 对象包含用户验证信息和用户权限信息，若用户验证失败，则弹出【密码错误】等对话框，并保持登录界面不变；如果验证通过，则关闭登录界面，弹出办公管理系统的客户端（参见图 7-1），并启动客户端 Socket 输入流监听线程 ClientSocketListener（参见代码 7-19）；最后，通过 ClientSocket 类（参见代码 7-20）的 addClientConnServer () 方法保存用户账号和 ClientConnServer 类对象。这样，当客户端与服务器端通信时，只需要从 ManageSocket 类中得到 ClientConnServer 类，再取得 Socket 就能获取输入输出流。至此，客户端已经准备好从 Socket 中读取来自服务器的消息对象。

【代码 7-15】 StartClient 类
```
package StartClientAndjudgeLogin;
public class StartClient {
    public static void main(String[] args) {
        new LoginUI();
    }
}
```

【代码 7-16】 LoginUI 类
```
package StartClientAndjudgeLogin;
import javax.swing.*;
import java.awt.event.*;
/* 登录界面 */
public class LoginUI extends JFrame implements ActionListener {
    private JFrame frame;
    private JButton enter;
```

```java
    private JPasswordField jPass;
    private JTextField jt_id;
    public LoginUI() {
        GUI();
    }
    public void GUI() {
        frame = new JFrame();
        //设置背景图片
        ImageIcon bg = new ImageIcon(
"src/image/imageStartClientAndjudgeLogin/background.png");
        JLabel label = new JLabel(bg);
        //背景图大小654*480
        label.setBounds(0, 0, bg.getIconWidth(), bg.getIconHeight());
        frame.getLayeredPane().add(label, new Integer(Integer.MIN_VALUE));
        JPanel jp = (JPanel) frame.getContentPane();
        jp.setOpaque(false);
        JPanel panel = new JPanel();
        panel.setOpaque(false);
        //文本输入框
        jt_id = new JTextField();
        jt_id.setBounds(250, 183, 155, 25);
        jPass = new JPasswordField();
        jPass.setBounds(250, 235, 155, 25);
        //登录按钮
        enter = new JButton();
        //鼠标经过的时候的图片
    ImageIcon entered = new ImageIcon( "src/image/imageStartClientAndjudgeLog-
in/button1.png");
        //初始图片
        ImageIcon released = new ImageIcon(
"src/image/imageStartClientAndjudgeLogin/button2.png");
        //按下去的图片
        ImageIcon pressed = new ImageIcon(
"src/image/imageStartClientAndjudgeLogin/button3.png");
        //透明化
        enter.setOpaque(false);
        enter.setIcon(released);//初始化
        enter.setBorderPainted(false);
        enter.setBorder(null);
        enter.setBounds(267, 290, 90, 32); //按钮大小,位置
        //按钮透明,这样用圆形按钮的时候也不会出现按钮边框
        enter.setContentAreaFilled(false);
```

```java
        //按下按钮时的图片
        enter.setPressedIcon(pressed);
        //经过时的图片
        enter.setRolloverIcon(entered);
        enter.addActionListener(this);
        frame.setSize(654,500);
        panel.setSize(654,500);
        panel.setLayout(null);
        panel.add(enter);
        panel.add(jt_id);
        panel.add(jPass);
        frame.add(panel);
        frame.setLocationRelativeTo(null);
        frame.setDefaultCloseOperation(JFrame.EXIT_ON_CLOSE);
        frame.setVisible(true);
    }

    public void actionPerformed(ActionEvent e){
        if(e.getSource() == enter){
            String pass = new String(jPass.getPassword());
            String id = jt_id.getText();
            try{
                BufferedReader br = new BufferedReader(new InputStreamReader(new FileInputStream("./server.cfg")));
                String serverIP = br.readLine();    //获取服务器IP地址
                new ClientAuthenticate(frame, id, pass, serverIP, 12345);
            }catch(FileNotFoundException e1){
                e1.printStackTrace();
            }catch(IOException e2){
                e2.printStackTrace();
            }
        }
    }
}
```

【代码7-17】 ClientAuthenticate 类

```java
package StartClientAndjudgeLogin;
import java.io.*;
import java.net.*;
import javax.swing.*;
import selectFunction.ClientMainUI;
import common.*;
/* 判断登录的信息及密码正确后的功能 */
```

```java
public class ClientAuthenticate {
    public ClientAuthenticate(JFrame frame, String id, String pass, String ip, int port) {
        try {
            //在软件部署前,必须设置正确的 IP
            Socket s = new Socket(ip, port);
            //发送 USER 包给 SERVER 端判断登录信息;
            ObjectOutputStream oos = new ObjectOutputStream(s.getOutputStream());
            User user = new User();
            user.setId(id);
            user.setPass(pass);
            oos.writeObject(user);
            //读取 USER 包看看验证的密码是否正确
            ObjectInputStream ois = new ObjectInputStream(s.getInputStream());
            User reuser = (User) ois.readObject();
            boolean judge_id = reuser.getType();
            if (judge_id == true) {//密码正确
                frame.dispose(); //登录界面消失
                String authority = reuser.getAuthority();
                new ClientMainUI(id, authority); //建立选择功能界面
                //用来做本 CLIENT 端所有的输入流(重要)。
                ClientConnServer ccs = new ClientConnServer(s);
                ccs.start(); //开启输入流线程
                /添加 SOCKET
                ManageSocket.addClientConnServer(id,ccs); /
            }
            else if (judge_id == false) {//密码错误
                JOptionPane.showMessageDialog(null, "密码错误");
            }
        } catch (Exception e) {
            e.printStackTrace();
        }
    }
}
```

【代码 7-18】 ClientMainUI 类

```java
package selectFunction;
import java.awt.event.*;
import javax.swing.*;
public class ClientMainUI extends JFrame implements ActionListener {
    private JFrame frame;
```

```java
        private JButton Checkannouncement; //查看公告
        private JButton Announcement; //发布公告
        private JButton chat; //聊天
        private JButton Signin; //签到
        private JButton Checkattendance; //查看考勤
        private JButton Certifiedstaff; //注册员工
        private JButton Deleteemployee; //删除员工
        private JButton Personalinformation;//查看个人信息
        private JButton Viewinginformation; //查看员工信息
        private JButton Editpinfo; //修改个人信息
        private String id, authority;
        private int flag = 0;//考勤要用到的
        public ClientMainUI(String id, String authority) {
            this.id = id;
            this.authority = authority;
            GUI();
        }
        public void buttonInit(JButton button, int x, int y, int w, int h, String s1, String s2, String s3){
            //鼠标经过时的图片
            ImageIcon entered = new ImageIcon("src/image/mainPage/" + s1 + ".png");
            //初始图片
            ImageIcon released = new ImageIcon("src/image/mainPage/" + s2 + ".png");
            //按下时的图片
            ImageIcon pressed = new ImageIcon("src/image/mainPage/" + s3 + ".png");
            button.setOpaque(false); //透明化
            button.setIcon(released); //初始化
            button.setBorderPainted(false);
            button.setBorder(null);
            button.setBounds(x, y, w, h); //按钮大小,位置
            //按钮透明,这样用圆形按钮的时候也不会出现按钮边框
            button.setContentAreaFilled(false);
            //按下按钮时的图片
            button.setPressedIcon(pressed);
            //经过时的图片
            button.setRolloverIcon(entered);
        }
        public void GUI(){ //功能选择界面
            this.setTitle("主界面");
```

```java
            ImageIcon bg = new ImageIcon("src/image/mainPage/background.png");
            JLabel label = new JLabel(bg);
            this.getLayeredPane().add(label, new Integer(Integer.MIN_VALUE));
            label.setBounds(0, 0, bg.getIconWidth(), bg.getIconHeight());
            this.setVisible(true);
            this.setSize(662, 513);
            //初始化查看公告按钮
            Checkannouncement = new JButton();
            buttonInit(Checkannouncement, 190, 170, 117, 37, "查看公告1", "查看公告2", "查看公告3");
            Announcement = new JButton();
            buttonInit(Announcement, 190, 220, 117, 37, "发布公告1", "发布公告2", "发布公告3");
            chat = new JButton();
            buttonInit(chat, 355, 170, 117, 37, "聊天1", "聊天2", "聊天3");
            Signin = new JButton();
            buttonInit(Signin, 20, 170, 117, 37, "签到1", "签到2", "签到3");
            Checkattendance = new JButton();
            buttonInit(Checkattendance, 20, 220, 117, 37, "查看考勤1", "查看考勤2", "查看考勤3");
            Certifiedstaff = new JButton();
            buttonInit(Certifiedstaff, 520, 270, 117, 37, "注册员工1", "注册员工2", "注册员工3");
            Deleteemployee = new JButton();
            buttonInit(Deleteemployee, 520, 320, 117, 37, "删除员工1", "删除员工2", "删除员工3");
            Personalinformation = new JButton();
            buttonInit(Personalinformation, 520, 170, 117, 37, "查看个人信息1", "查看个人信息2", "查看个人信息3");
            Viewinginformation = new JButton();
            buttonInit(Viewinginformation, 520, 370, 117, 37, "查看员工信息1", "查看员工信息2", "查看员工信息3");
            Editpinfo = new JButton();
            buttonInit(Editpinfo, 520, 220, 117, 37, "修改个人信息1", "修改个人信息2", "修改个人信息3");
            //添加监听器
            Checkannouncement.addActionListener(this);
            Announcement.addActionListener(this);
            chat.addActionListener(this);
            Signin.addActionListener(this);
            Checkattendance.addActionListener(this);
            Certifiedstaff.addActionListener(this);
```

```java
        Deleteemployee.addActionListener(this);
        Personalinformation.addActionListener(this);
        Viewinginformation.addActionListener(this);
        Editpinfo.addActionListener(this);
        JPanel jp = (JPanel) this.getContentPane();
        jp.setOpaque(false);
        JPanel panel = new JPanel();
        panel.setSize(662,513);
        panel.setLayout(null);
        panel.setOpaque(false);
        this.add(panel);
        this.setSize(662,513);
        panel.add(Checkannouncement);
        panel.add(Announcement);
        panel.add(chat);
        panel.add(Signin);
        panel.add(Checkattendance);
        panel.add(Certifiedstaff);
        panel.add(Deleteemployee);
        panel.add(Personalinformation);
        panel.add(Viewinginformation);
        panel.add(Editpinfo);
        this.setLocationRelativeTo(null);
        this.setDefaultCloseOperation(JFrame.EXIT_ON_CLOSE);
        this.setVisible(true);
    }
    public boolean judgeAuthority(String authority){
        boolean result = false;
        if (authority.equals("经理")){
          result = true;
        }
        else{
          JOptionPane.showMessageDialog(null,"权限不足");
        }
        return result;
    }
    public void actionPerformed(ActionEvent e){
        if (e.getSource() == Checkannouncement){
            //公告的按钮
            //new browser(id,authority);
        }
        else if (e.getSource() == Announcement){
```

```java
            if (judgeAuthority(authority)) {
                //new publicSend(id);
            }
        }
        else if (e.getSource() == Certifiedstaff) {
            //员工信息的按钮
            if (judgeAuthority(authority)) {
                //new EnrollUI(id);
            }
        }
        else if (e.getSource() == Deleteemployee) {
            if (judgeAuthority(authority)) {
                //new deleteStaffUI(id);
            }
        }
            else if (e.getSource() == Personalinformation) {
                //new FindPersonInformation(id);
            }
        else if (e.getSource() == Viewinginformation) {
            if (judgeAuthority(authority)) {
                //new findAllStaffInformation(id);
            }
        }
        else if (e.getSource() == Editpinfo) {
            if (judgeAuthority(authority)) {
                //new ModifyPersonMs(id);
            }
        }
        else if (e.getSource() == chat) {
            //聊天的按钮
            //chatFunction chat1 = new chatFunction(id, this);
        }
        else if (e.getSource() == Signin) {
            //考勤的按钮
            if (flag == 0) {
                flag = 1;
                //new attendFucation(id).writeIn();
            }
            else if (flag == 1) {
                JOptionPane.showMessageDialog(null, "你今天已经签到");
            }
        }
```

```
       else if (e.getSource() = = Checkattendance) {
           if (judgeAuthority(authority)) {
               //new init(id);
           }
       }
     }
   }
}
```

【代码 7 – 19】ClientSocketListener 类

```
package ManageClientConnServer;
import java.io.ObjectInputStream;
import java.net. *;
import common.Message;
/* 专门用来读取客户端 Socket 中的数据 */
public class ClientSocketListener extends Thread {
    private Socket s;
    public ClientSocketListener(Socket s) { this.s = s; }
    public Socket getS() { return s; }
    public void run() {
      while (true) {
          try {
           ObjectInputStream ois = new ObjectInputStream(s.getInputStream
());
              Message ms = (Message) ois.readObject();
              //后续与服务器端交互信息的代码加在这里;
           }catch (Exception e) {
           break;
           }
        }
     }
}
```

【代码 7 – 20】ClientSocket 类

```
package ManageClientConnServer;
import java.util.HashMap;
public class ClientSocket {
     private static HashMap Socket_Hash = new HashMap < String , ClientSock-
etListener >();
     public static void addClientConnServer(String id , ClientSocketListener
ccs ) {
         Socket_Hash.put( id, ccs);
     }
     public static ClientSocketListener getClientConnServer(String id) {
```

```
        return (ClientSocketListener)Socket_Hash.get(id);
    }
}
```

7.3.2 增量7-2：信息管理模块

增量7-2需要实现用例11至用例16，包括查看个人信息、修改个人信息、注册员工、删除员工、查看员工信息共5个功能，具体的功能流程图请参考图7-8、图7-10、图7-12、图7-14、图7-16。

1. 查看个人信息

根据图7-35序列图实现查看个人信息功能。用户单击【查询个人信息】按钮，功能选择界面会创建FindPersonInformation（参见代码7-22）对象，并使用Socket向服务器端发出个人信息查询Message，消息类型为3，因为服务器端返回消息需要一段时间才能传送回给客户端，所以接下来客户端通过sleep（）方法等待100毫秒；服务器端的客户服务线程收到此消息后，通过JDBC从数据库中读取用户信息，把信息保存在一个Message（参见代码7-13）对象中返回给客户端；客户端Socket监听器ClientSocketListener（参见代码7-24）把从服务器端收到的消息保存在个人信息管理类ManageMsPerson（参见代码7-25）中；客户端主线程休眠结束，从个人信息管理类中读取出信息并显示在FindPersonInformation界面上。

ManageMsPerson类用于在客户端保存用户个人信息，用户个人信息被存储在一个HashMap集合对象中，访问者可以通过用户ID号读取相应的用户信息（可采用ArrayList方式存储）。

图7-35 查询个人信息序列

根据图7-35所示序列图，依次实现以下两个步骤。
（1）客户端：实现功能选择界面对【查看个人信息】按钮的事件处理。
首先，为客户端功能选择界面（ClientMainUI，代码7-18）增加【查看个人信息】

按钮的事件处理代码（参见代码7-21），当【查看个人信息】按钮事件触发时，创建并显示员工信息界面（参见代码7-22），员工信息界面类 FindPersonInformation 继承自 FindAndModifyCommonUI 类（参见代码7-23），员工个人信息界面见图7-7。

【代码7-21】修改 ClientMainUI 类，增加"查看个人信息"的事件处理。

```java
public void actionPerformed(ActionEvent e){
...
        if (e.getSource() == Personalinformation){
        new FindPersonInformation(id);
    }
...
}
```

【代码7-22】FindPersonInformation 类

```java
package staffMessFunction;
import javax.swing.ImageIcon;
import javax.swing.JLabel;
public class FindPersonInformation extends FindAndModifyCommonUI{
    private ImageIcon image;
    private JLabel labelImage;
    public FindPersonInformation(String id) {
        super(id);
        image = new ImageIcon("src/image/staffManage/公告背景.png");
        labelImage = new JLabel(image);
        this.getLayeredPane().add(labelImage, new Integer(Integer.MIN_VALUE));
        labelImage.setBounds ( 0, 0, image.getIconWidth ( ), image.getIconHeight());
        this.setTitle("查看个人信息");
    }
}
```

【代码7-23】FindAndModifyCommonUI 类

```java
package staffMessFunction;
import java.io.IOException;
import java.io.ObjectOutputStream;
import java.util.ArrayList;
import javax.swing.*;
import ManageClientConnServer.ClientSocket;
import ManageClientConnServer.ManageMsPerson;
import common.Message;
public class FindAndModifyCommonUI extends JFrame{
    private JButton JB_confirmEnroll, JB_deleteMess;
    protected JTextField JT_income, JT_id, JT_authority, JT_age, JT_telNum,
```

```java
JT_Address, JT_sex, JT_QQnum;
    protected String id;
    protected ArrayList msList;
    private JPanel root;
    public FindAndModifyCommonUI(String id) {
        this.id = id;
        Message ms = new Message();//消息类型3 表示个人信息查询消息
        ms.setMesstype(3);
        ms.setSender(id);
        ms.setGetter(id);
        try {
            ObjectOutputStream oos = new ObjectOutputStream(new ClientSocket().getClientConnServer(id).getS().getOutputStream());
            oos.writeObject(ms);
            Thread.sleep(100); //等待服务器查询结果
        } catch (IOException e) {
            e.printStackTrace();
        } catch (InterruptedException e) {
            e.printStackTrace();
        }
        msList = ManageMsPerson.getArrayList(id);
        if(msList != null)
            GUI(msList);
    }
    public void GUI(ArrayList msList) {
        this.setLayout(null);
        root = (JPanel) this.getContentPane();
        root.setOpaque(false);
        JT_id = new JTextField(msList.get(0) + "");
        JT_authority = new JTextField(msList.get(1) + "");
        JT_income = new JTextField(msList.get(3) + "");
        JT_age = new JTextField(msList.get(4) + "");
        JT_sex = new JTextField(msList.get(5) + "");
        JT_Address = new JTextField(msList.get(6) + "");
        JT_telNum = new JTextField(msList.get(7) + "");
        JT_QQnum = new JTextField(msList.get(8) + "");
        JT_id.setEditable(false);
        JT_authority.setEditable(false);
        JT_income.setEditable(false);
        JT_age.setEditable(false);
        JT_telNum.setEditable(false);
        JT_Address.setEditable(false);
```

```java
JT_sex.setEditable(false);
JT_QQnum.setEditable(false);
JLabel JL_id = new JLabel("姓 名 :");
JLabel JL_authority = new JLabel("权 限 :");
JLabel JL_income = new JLabel("收 入 :");
JLabel JL_age = new JLabel("年 龄 :");
JLabel JL_sex = new JLabel("性 别 :");
JLabel JL_telNum = new JLabel("联系电话:");
JLabel JL_Address = new JLabel("家庭住址:");
JLabel JL_QQnum = new JLabel("Q Q 号码:");
JB_confirmEnroll = new JButton("确定");
JB_deleteMess = new JButton("清空信息");
int x = 265, y = 38;
JT_id.setBounds(x, y + 30, 150, 20);
JT_authority.setBounds(x, 2 * y + 30, 150, 20);
JT_income.setBounds(x, 3 * y + 30, 150, 20);
JT_age.setBounds(x, 4 * y + 30, 150, 20);
JT_telNum.setBounds(x, 5 * y + 30, 150, 20);
JT_Address.setBounds(x, 6 * y + 30, 150, 20);
JT_sex.setBounds(x, 7 * y + 30, 150, 20);
JT_QQnum.setBounds(x, 8 * y + 30, 150, 20);
JB_confirmEnroll.setBounds(150, 380, 100, 25);
JB_deleteMess.setBounds(270, 380, 100, 25);
this.add(JT_id);
this.add(JT_authority);
this.add(JT_income);
this.add(JT_age);
this.add(JT_telNum);
this.add(JT_Address);
this.add(JT_sex);
this.add(JT_QQnum);
this.add(JL_id);
this.add(JL_authority);
this.add(JL_income);
this.add(JL_age);
this.add(JL_telNum);
this.add(JL_Address);
this.add(JL_sex);
this.add(JL_QQnum);
this.setSize(662, 513);
this.setLocation(450, 150);
```

```
            this.setVisible(true);
    }
}
```

【代码 7-24】 修改 ClientSocketListener 类，增加对用户信息接收的处理

```
...
public void run() {
    ...
    //后续与服务器端交互信息的代码加在这里；
    if (ms.getMesstype() == 3)  //员工信息
    {
        String id = ms.getGetter();
        ArrayList list = ms.getList();
        ManageMsPerson.addArrayList(id, list);
    }
    ...
}
```

【代码 7-25】 ManageMsPerson 类

```
package ManageClientConnServer;
import java.util.ArrayList;
import java.util.HashMap;
public class ManageMsPerson {
    private static HashMap msPerson_Hash = new HashMap<String, ArrayList>();
    public static void addArrayList(String id, ArrayList msPersonList) {
        msPerson_Hash.put(id, msPersonList);
    }
    public static ArrayList getArrayList(String id) {
        return (ArrayList)msPerson_Hash.get(id);
    }
}
```

服务器端：实现功能选择界面对【查看个人信息】按钮的事件处理。客户服务线程 ClientHandler 每收到一个消息，首先判断消息类型，如果收到个人信息查询消息（Message 类型 3），则通过 JDBC 编程接口从数据库中查询指定用户 ID 的信息，并以消息的形式返回给客户端 Socket 监听器。代码 7-26 实现了上述处理过程，代码 7-27 是 JDBC 查询功能的实现。

【代码 7-26】 修改 ClientHandler 类，实现对消息类型 3 的处理

```
public void run() {
    ...
    ObjectInputStream ois = new ObjectInputStream(s.getInputStream());
    Message ms = (Message) ois.readObject();
    //后续与客户端通信的代码加在这里；
```

```java
        if (ms.getMesstype() = = 3) {//申请员工信息
            String id = ms.getSender();
            ArrayList personMs = new JDBC().back_StaffPersonMessage(id);
            ObjectOutputStream oos = new ObjectOutputStream (ClientHandlerManager.getClientHandler(ms.getGetter()).getS().getOutputStream());
            Message three_backMs = new Message();
            hree_backMs.setSender(id);
            three_backMs.setGetter(ms.getGetter());
            three_backMs.setMesstype(3);
            three_backMs.setList(personMs);
            oos.writeObject(three_backMs);
        }
    ...
}
```

【代码 7 - 27】修改 JDBC 类，增加个人信息查询方法

```java
public ArrayList back_StaffPersonMessage(String id) {//返回员工信息
    ArrayList PersonMs = new ArrayList();
    try{
        Class.forName(driver);
        conn = DriverManager.getConnection(url, userName, sqlpassword);
        stmt = conn.createStatement();
        String sqlSelect = "SELECT id ,authority,pass,income,age,sex,address,telpNum,QQNum" + "FROM userInfo" + "Where id ='" + id + "'";
        rs = stmt.executeQuery(sqlSelect);
        while (rs.next()) {
            PersonMs.add(rs.getString(1));
        PersonMs.add(rs.getString(2));
        PersonMs.add(rs.getString(3));
        PersonMs.add(rs.getInt(4));
        PersonMs.add(rs.getInt(5));
        PersonMs.add(rs.getString(6));
        PersonMs.add(rs.getString(7));
        PersonMs.add(rs.getString(8));
        PersonMs.add(rs.getString(9));
        }
        rs.close();
        stmt.close();
        conn.close();
    }catch (ClassNotFoundException e) {
        e.printStackTrace();
    }catch (SQLException e) {
```

```
            e.printStackTrace();
        }
        return PersonMs;
}
```

2. 修改个人信息

修改个人信息与查看个人信息的思路是相同的，都是先从服务器端的数据库读取员工名单，当单击【修改】按钮（个人信息修改界面请参见图 7-11）后，可在相应位置对个人信息进行修改并单击【确认修改】后，客户端会对前后信息进行比对，若有不同，则向服务器端发送类型为 4 的消息，服务器收到消息后通过 JDBC 访问数据库完成用户信息的更新。具体实现步骤如下：

（1）客户端：实现功能选择界面对【修改个人信息】按钮的事件处理。当用户单击【修改个人信息】按钮（代码 7-28）时，功能选择界面会检查用户权限，如果是经理权限（"mgr"），则弹出用户信息修改界面（参见代码 7-29），否则提示用户权限不足。

用户信息修改界面在初始化后，允许用户更改用户的各项信息，如果用户确认修改，则把被修改的用户信息以消息（消息类型为 4）对象的方式发送到服务器进行修改。

【代码 7-28】修改 ClientMainUI 类，增加【修改个人信息】按钮的事件处理

```
public void actionPerformed(ActionEvent e) {
...
    else if (e.getSource() = = Editpinfo) {
        if (judgeAuthority(authority)) {
            new ModifyPersonMs(id);
        }
    }
...
}
```

（2）客户端：实现【修改个人信息】界面和事件处理。

【代码 7-29】ModifyPersonInformation 类

```
package staffMessFunction;
import java.awt.event.ActionEvent;
import java.awt.event.ActionListener;
import java.io.IOException;
import java.io.ObjectOutputStream;
import java.util.ArrayList;
import javax.swing.*;
import ManageClientConnServer.ClientSocket;
import common.Message;
public class ModifyPersonInformation extends FindAndModifyCommonUI imple-
ments ActionListener{
```

```java
        private JTextField JT_pass;
        private String id;
        private JButton jbModify;
        private JButton jbConfirm;
        private ImageIcon image;
        private JLabel labelImage;
        public ModifyPersonInformation(String id){
            super(id);
            this.id = id;
            this.setTitle("修改个人信息");
            GUI();
        }
        public void ModifyMsgSent(ArrayList list){ //发送修改信息
            Message ms = new Message();
            ms.setList(list);
            ms.setMesstype(4);
            ms.setSender(id);
            try {
                ObjectOutputStream oos = new ObjectOutputStream(new ClientSocket().getClientConnServer(id).getS().getOutputStream());
                oos.writeObject(ms);
            }catch (IOException e1){
                e1.printStackTrace();
            }
        }
    public void GUI(){
            image = new ImageIcon("src/image/staffManage/公告背景.png");
            labelImage = new JLabel(image);
            this.getLayeredPane().add(labelImage, new Integer(Integer.MIN_VALUE));
            labelImage.setBounds ( 0, 0, image.getIconWidth ( ), image.getIconHeight());
            JT_pass = new JTextField();
            JLabel JL_pass = new JLabel("密 码:");
            JT_pass.setBounds(200 + 65, 9 * 38 + 30, 150, 20);
            JL_pass.setBounds(185, 9 * 38 + 30, 150, 30);
            JT_pass.setEditable(false);
            JT_pass.setText(msList.get(2) + "");
            this.add(JT_pass);
            this.add(JL_pass);
            jbModify = new JButton("修改信息");
            jbModify.setBounds(325, 405, 100, 30);
```

```java
            //鼠标经过时的图片
            ImageIcon entered = new ImageIcon("src/image/staffManage/修改2.png");
            //初始图片
            ImageIcon released = new ImageIcon("src/image/staffManage/修改3.png");
            //按下时的图片
            ImageIcon pressed = new ImageIcon("src/image/staffManage/修改1.png");
            jbModify.setOpaque(false); //透明化
            jbModify.setIcon(released); //初始化
            jbModify.setBorderPainted(false);
            jbModify.setBorder(null);
            jbModify.setBounds(325,405,100,30); //按钮大小,位置
            //按钮透明,这样用圆形按钮的时候也不会出现按钮边框
            jbModify.setContentAreaFilled(false);
            jbModify.setPressedIcon(pressed); //按下按钮时的图片
            jbModify.setRolloverIcon(entered); //经过时的图片
            jbConfirm = new JButton("确认修改");
            jbConfirm.setBounds(205,405,100,30);
            ImageIcon entered2 = new ImageIcon("src/image/staffManage/确认修改2.png");
            ImageIcon released2 = new ImageIcon("src/image/staffManage/确认修改3.png");
            ImageIcon pressed2 = new ImageIcon("src/image/staffManage/确认修改1.png");
            jbConfirm.setOpaque(false); //透明化
            jbConfirm.setIcon(released2); //初始化
            jbConfirm.setBorderPainted(false);
            jbConfirm.setBorder(null);
            jbConfirm.setBounds(205,405,100,30); //按钮大小,位置
            jbConfirm.setContentAreaFilled(false);
            jbConfirm.setPressedIcon(pressed2);
            jbConfirm.setRolloverIcon(entered2);
            this.add(jbModify);
            this.add(jbConfirm);
            jbModify.addActionListener(this);
            jbConfirm.addActionListener(this);
    }
            @Override
            public void actionPerformed(ActionEvent e) {
            JTextField[] jts = { JT_id, JT_authority, JT_pass, JT_income, JT_age,
```

```
JT_sex, JT_Address, JT_telNum, JT_QQnum};
        String[] jt_labels = {"id", "authority", "pass", "income", "age", "sex", "address", "telpNum", "QQNum"};
          if(e.getSource() = = jbModify){
            for(int i = 0; i < jts.length; i + +)
              jts[i].setEditable(true);
    }
      else if(e.getSource() = = jbConfirm){
        for(int i = 0; i < jts.length; i + +)
        jts[i].setEditable(false);
        for(int i = 0; i < jts.length; i + +){
          if((i = = 3 || i = = 4) && (msList.get(i) + "").equals(jts[i].getText())){
            ArrayList list = new ArrayList();
            list.add(jt_labels[i]);
            list.add(jts[i].getText());
            ModifyMsgSent(list);
          }
          else if(!((msList.get(i) + "").equals(jts[i].getText()))){
            ArrayList list = new ArrayList();
            list.add(jt_labels[i]);
            list.add(jts[i].getText());
            ModifyMsgSent(list);
          }
          System.out.println(msList.get(i) + " " + jts[i].getText());
        }
      }
    }
  }
```

（3）服务器端：实现个人信息修改。首选，在客户端服务线程中实现对 Message 类型为 4 消息对象的处理（参见代码 7 - 30），然后实现数据库编程接口 JDBC 类对数据库中用户信息更新的功能，这需要在 JDBC 类中增加一个 ModifyPersonInfo( ) 方法（参见代码 7 - 31）。

【代码 7 - 30】修改 ClientHandler 类，增加对消息类型 4 的处理

```
public void run(){
...
    else if(ms.getMesstype() = = 4){ //接受修改信息
      ArrayList list = ms.getList();
      String id = ms.getSender();
      new JDBC().modifyPersonMS(list, id);
    }
    ...
}
```

【代码7-31】修改 JDBC 类,增加用户个人信息更新的 modifyPersonInfo()方法:
...
```
public void modifyPersonInfo(ArrayList list, String id) {//修改个人信息
    try{
        Class.forName(driver);
        conn = DriverManager.getConnection(url, userName, sqlpassword);
        stmt = conn.createStatement();
        String sql = "update userInfo" + "set" + list.get(0) + "=" + "'" +
list.get(1) + "'" + "where id ='" + id + "'";
        stmt.executeUpdate(sql);
    } catch (ClassNotFoundException e){
        e.printStackTrace();
    }catch (SQLException e){
        e.printStackTrace();
    }
}
```
...

3. 注册员工

(经理)注册员工功能比较简单,首先由具有经理极限的用户创建注册界面(如图7-13所示);然后在输入框录入要注册的信息,以 Message 对象(使用消息类型1)记录待注册员工信息;再把此消息对象发送给服务器。服务器收到类型为1的 Message 对象后,在数据库增加员工记录即可。具体实现步骤如下:

(1)客户端:实现注册员工界面。首先,在功能选择界面(ClientMainUI)增加对【注册员工】按钮的事件处理(参见代码7-32),然后创建注册员工界面(参见代码7-33),当经理输入员工信息后并单击【注册】按钮,注册员工界面就会发送一个类型为1的 Message 对象给服务器端的客户服务线程。此消息中包含了待注册员工的信息。

【代码7-32】修改 ClientMainUI 类,增加【注册员工】按钮的事件处理
```
public void actionPerformed(ActionEvent e) {
    ...
    else if(e.getSource() == Certifiedstaff){//注册员工
        if(judgeAuthority(authority)){
            new EnrollUI(id);
        }
    }
    ...
}
```

【代码7-33】EnrollGUI 类
```
package staffMessFunction;
import java.awt.event.*
```

```java
import java.io.IOException;
import java.io.ObjectOutputStream;
import java.net.Socket;
import javax.swing.*;
import ManageClientConnServer.ClientSocket;
import common.Message;
    public class EnrollGUI extends JFrame implements ActionListener{
    private JButton JB_confirmEnroll, JB_deleteMess;
    private JComboBox JT_authority;
    private JTextField JT_income, JT_id, JT_age, JT_telNum;
    private JTextField JT_Address, JT_sex, JT_QQnum;
    private JPasswordField JP_pass;
    private String id;
    private ImageIcon image;
    private JLabel labelImage;
    private JPanel root;
    public EnrollGUI(String id){
        image = new ImageIcon("src/image/staffManage/公告背景.png");
        labelImage = new JLabel(image);
        this.getLayeredPane().add(labelImage, new Integer(Integer.MIN_VALUE));
        labelImage.setBounds(0, 0, image.getIconWidth(), image.getIconHeight());
        root = (JPanel) this.getContentPane();
        root.setOpaque(false);
        this.id = id;
        this.setLayout(null);
        this.setSize(662, 513);
        this.setTitle("注册");
        this.setLocation(400, 140);
        JT_id = new JTextField("");
        String[] authority = { "mgr", "user" };
        JT_authority = new JComboBox(authority);
        JP_pass = new JPasswordField("");
        JT_income = new JTextField("");
        JT_age = new JTextField("");
        JT_telNum = new JTextField("");
        JT_Address = new JTextField("");
        JT_sex = new JTextField("");
        JT_QQnum = new JTextField("");
        JLabel JL_id = new JLabel("姓名:");
        JLabel JL_authority = new JLabel("权限:");
```

```java
JLabel JL_pass = new JLabel("密 码 :");
JLabel JL_income = new JLabel("收 入 :");
JLabel JL_age = new JLabel("年 龄 :");
JLabel JL_sex = new JLabel("性 别 :");
JLabel JL_telNum = new JLabel("联系电话:");
JLabel JL_Address = new JLabel("家庭住址:");
JLabel JL_QQnum = new JLabel("Q Q 号码:");
JB_confirmEnroll = new JButton();
//鼠标经过时的图片
    ImageIcon entered = new ImageIcon ( " src/image/staffManage/确定xz.png");
//初始图片
    ImageIcon released = new ImageIcon ( " src/image/staffManage/确定yx.png");
//按下时的图片
    ImageIcon pressed = new ImageIcon ( " src/image/staffManage/确定ax.png");
JB_confirmEnroll.setOpaque(false);  //透明化
JB_confirmEnroll.setIcon(released);  //初始化
JB_confirmEnroll.setBorderPainted(false);
JB_confirmEnroll.setBorder(null);
JB_confirmEnroll.setBounds(285,360,92,33);  //按钮大小,位置
JB_confirmEnroll.setContentAreaFilled(false);  //按钮透明
JB_confirmEnroll.setPressedIcon(pressed);  //鼠标按下按钮时的图片
JB_confirmEnroll.setRolloverIcon(entered);  //鼠标经过时的图片
JB_confirmEnroll.addActionListener(this);
JT_id.setBounds(200 + 65,60 + 20,150,20);
JT_authority.setBounds(200 + 65,90 + 20,150,20);
JP_pass.setBounds(200 + 65,120 + 20,150,20);
JT_income.setBounds(200 + 65,150 + 20,150,20);
JT_age.setBounds(200 + 65,180 + 20,150,20);
JT_sex.setBounds(200 + 65,210 + 20,150,20);
JT_Address.setBounds(200 + 65,240 + 20,150,20);
JT_telNum.setBounds(200 + 65,270 + 20,150,20);
JT_QQnum.setBounds(200 + 65,300 + 20,150,20);
JL_id.setBounds(120 + 65,80,150,20);
JL_authority.setBounds(120 + 65,110,150,20);
JL_pass.setBounds(120 + 65,140,150,20);
JL_income.setBounds(120 + 65,170,150,20);
JL_age.setBounds(120 + 65,200,150,20);
JL_sex.setBounds(120 + 65,230,150,20);
JL_Address.setBounds(120 + 65,260,150,20);
```

```java
            JL_telNum.setBounds(120 + 65, 290, 150, 20);
            JL_QQnum.setBounds(120 + 65, 320, 150, 20);
            This.add(JB_confirmEnroll);
            this.add(JT_id);
            this.add(JT_authority);
            this.add(JP_pass);
            this.add(JT_income);
            this.add(JT_age);
            this.add(JT_telNum);
            this.add(JT_Address);
            this.add(JT_sex);
            this.add(JT_QQnum);
            this.add(JL_id);
            this.add(JL_authority);
            this.add(JL_pass);
            this.add(JL_income);
            this.add(JL_age);
            this.add(JL_telNum);
            this.add(JL_Address);
            this.add(JL_sex);
            this.add(JL_QQnum);
            JB_confirmEnroll.addActionListener(this);
            this.setVisible(true);
        }
        @Override
        public void actionPerformed(ActionEvent e) {
            if (e.getSource() = = JB_confirmEnroll){
                Message ms = new Message();
                ms.setMesstype(1);
                String content = JT_id.getText() + " " + JT_authority.getSelectedItem()
                + " " + new String(JP_pass.getPassword()) + " "
                + JT_income.getText() + " " + JT_age.getText() + " "
                + JT_sex.getText() + " " + JT_Address.getText() + " "
                + JT_telNum.getText() + " " + JT_QQnum.getText();
                ms.setContent(content);
                Socket s = ClientSocket.getClientConnServer(id).getS();
                try {
                ObjectOutputStream oos = new ObjectOutputStream(s.getOutputStream());
                oos.writeObject(ms);
            }catch (IOException e1) {
```

```
            e1.printStackTrace();
          }
        this.dispose();
       }
    }
}
```

(2) 服务器端：增加新用户记录。客户服务线程 ClientHandler 在接收到类型为 1 的 Message 对象（参见代码 7-34）时，调用 JDBC 类的 enroll() 方法（需要在 JDBC 类中增加，参见代码 7-35），enroll() 方法通过向数据库的用户表中插入一个新的员工记录，来完成新用户注册过程。

【代码 7-34】修改 ClientHandler 类，增加对类型为 1 的 Message 处理

```
public void run(){
    ...
    if(ms.getMesstype() == 1){  //接受修改信息
       String[] cont = ms.getContent().split(" ");
       new JDBC().enroll(cont);
    }
    else if(ms.getMesstype() == 3)  //申请员工信息
    ...
}
```

【代码 7-35】修改 JDBC 类，增加新用户注册的 enroll() 方法

```
public void enroll(String[] cont){
    try{
        Class.forName(driver);
        conn = DriverManager.getConnection(url, userName, sqlpassword);
        stmt = conn.createStatement();
        String sql = "insert into userinfo" + "values(" + "'" + cont[0]
+ "','"
             + cont[1] + "','"
             + cont[2] + "'," + Integer.parseInt(cont[3]) + ","
             + Integer.parseInt(cont[4]) + ",'" + cont[5] + "','"
             + cont[6] + "','" + cont[7] + "','" + cont[8] + "')";
      stmt.executeUpdate(sql);
    }catch(ClassNotFoundException e){
      e.printStackTrace();
    }catch(SQLException e){
      e.printStackTrace();
    }
  }
}
```

4. 删除员工

首先由具有经理权限的用户向服务器发起读取员工名单请求的消息（Message 类型

为 2）。服务器从数据库中查找用户列表后，将类型为 2 的 Message 对象反馈给客户端，每个用户信息发送一个 Message 对象。客户端收到这些消息后，把用户信息显示在员工列表界面（如图 7 – 15 所示）。在删除员工界面上，经理用户可以通过单击删除界面左侧的【员工名单】中的用户名来删除用户，单击后，待删除的用户名就出现在界面右侧的【删除名单】中，再单击【删除】按钮即可完成删除操作。具体实现步骤如下：

（1）客户端：实现删除员工界面和事件处理。首先实现客户端功能选择界面中【删除员工】按钮的事件处理功能（参见代码 7 – 36）。

【代码 7 – 36】修改 ClientMainUI 类，增加对【删除员工】按钮的事件处理

```
public void actionPerformed(ActionEvent e){
  …
    else if(e.getSource() = = Deleteemployee){
      if(judgeAuthority(authority)){
        new DeleteStaffUI(id);
      }
    }
  …
}
```

然后，实现员工删除界面（DeleteStaffUI 类：参见代码 7 – 37）以及事件处理。由于用户数较多，所以 DeleteStaffUI 使用分页技术分页显示用户列表。

分页做法：使用 Page 变量和 MaxPage 变量来实现分页效果，Page 变量是当前页数，MaxPage 变量是最大页数，每页有 5 个 JLabel，用于显示 5 个员工信息，当我们单击【下一页】时需从 ArrayList 中读取相应位置的数据并对 JLabel 赋值，同时更新当前页数 Page 变量。同理，【上一页】的功能也可类似实现。

选中某条员工记录后，单击【删除员工】按钮删除服务器中选定的员工记录。在 DeleteStaffUI 监听到来自【删除员工】按钮的事件［DeleteStaffUI 类的 actionPerformed()方法］时，发送一个删除员工的 Message 对象（Message 类型为 5）给服务器。如果经理用户同时选中了多个员工记录，则只需要发送删除员工的 Message 对象给服务器即可，通过 DeleteStaffUI 类更新删除界面上显示的员工列表。

最后，实现客户端 Socket 监听器 ClientSocketListener 对消息类型为 2 的用户列表的处理（参见代码 7 – 38）。

【代码 7 – 37】DeleteStaffUI 类
```
package staffMessFunction;
import java.awt.*;
import java.io.IOException;
import java.io.ObjectOutputStream;
import java.util.ArrayList;
import javax.swing.*;
import ManageClientConnServer.ClientSocket;
import ManageClientConnServer.ManageStaffList;
```

```java
import common.Message;
public class DeleteStaffUI extends JFrame implements MouseListener, ActionListener{
    private JButton JB_delete;
    private JLabel[] label;
    private String id;
    private ArrayList<String> nameList;
    private int page = 1, maxPage;
    private JLabel JL_nextPage, JL_prePage;
    public DeleteStaffUI(String id){
        nameList = new ArrayList();
        this.id = id;
        this.getContentPane().setBackground(new Color(1,100,124));
        try {
            Message ms = new Message();
            ms.setMesstype(2);
            ms.setSender(id);
            ObjectOutputStream oos = new ObjectOutputStream(ClientSocket.getClientConnServer(id).getS().getOutputStream());
            oos.writeObject(ms);
            Thread.sleep(100); //等待服务器返回用户列表
            String[] list = ManageStaffList.getStringList(id);
            for(int i = 0; i < list.length; i++){
                nameList.add(list[i]);
            }
            UpdateMaxPage();
            GUI(nameList);
        }catch(Exception e) {
            e.printStackTrace();
        }
    }
    public void GUI(ArrayList list){
        this.setLayout(null);
        this.setTitle("删除员工列表");
        label = new JLabel[list.size()];
        JLabel JL_nameList = new JLabel("员工名单");
        JLabel JL_Label = new JLabel("删除名单");
        JL_prePage = new JLabel("上一页");
        JL_prePage.setBounds(120,300,50,30);
        JL_nextPage = new JLabel("下一页");
        JL_nextPage.setBounds(180,300,50,30);
        JL_prePage.addMouseListener(this);
```

```java
        JL_nextPage.addMouseListener(this);
        JL_nameList.setForeground(new Color(187,187,187));
        JL_Label.setForeground(new Color(187,187,187));
        JL_nameList.setBounds(30,20,80,20);
        JL_Label.setBounds(150,20,80,20);
        this.add(JL_nameList);
        this.add(JL_Label);
        this.add(JL_prePage);
        this.add(JL_nextPage);
        for(int i = (page -1) * 5; i < (page -1) * 5 + 5; i++){
            if(i < nameList.size()){
                int k = i % 5;
                label[k] = new JLabel(nameList.get(i).toString());
                label[k].setBounds(30,50 + k * 30,80,20);
                label[k].setForeground(new Color(187,187,187));
                label[k].addMouseListener(this);
                this.add(label[k]);
            }
            else
                break;
        }
        JB_delete = new JButton();
        ImageIcon entered = new ImageIcon("src/image/staffManage/删除2.png");
        ImageIcon released = new ImageIcon("src/image/staffManage/删除3.png");
        ImageIcon pressed = new ImageIcon("src/image/staffManage/删除1.png");
        JB_delete.setOpaque(false);//透明化
        JB_delete.setIcon(released);//初始化
        JB_delete.setBorderPainted(false);
        JB_delete.setBorder(null);
        JB_delete.setBounds(20,300,92,33); //按钮大小,位置
        JB_delete.setContentAreaFilled(false);//按钮透明
        JB_delete.setPressedIcon(pressed);//按下按钮时的图片
        JB_delete.setRolloverIcon(entered); //经过时的图片
        JB_delete.addActionListener(this);
        this.add(JB_delete);
        this.setSize(250,400);
        this.setLocation(500,300);
        this.setVisible(true);
    }
    @Override
    public void actionPerformed(ActionEvent e){
        if(e.getSource() == JB_delete){
```

```java
                    ArrayList deleteID = new ArrayList();
                    ArrayList <Integer> deleteLocation = new ArrayList();
                    for(int i = 0; i < 5; i + +){
                        if(i < nameList.size()){
                            if(label[i].getForeground().equals(Color.BLUE)){
                                deleteID.add(label[i].getText());
                                deleteLocation.add((page -1) * 5 + i);
                                label[i].setForeground(new Color(187, 187, 187));
                                Rectangle rs = label[i].getBounds();
                                rs.x - = 130;
                                label[i].setBounds(rs);
                                label[i].setText("");
                            }
                        }
                        else
                            break;
                    }
                    int deleteNum = deleteLocation.size();
                    for(int k = 0; k < deleteNum; k + +){
                        int i = deleteLocation.get(deleteLocation.size() -1);
                        nameList.remove(i);
                        UpdateMaxPage();
                        deleteLocation.remove(deleteLocation.size() -1);
                    }
                    this.repaint();
                    try{ //发送删除名单
                        ObjectOutputStream oos = new ObjectOutputStream(ClientSocket.getClientConnServer(id).getS().getOutputStream());
                        Message deleteMs = new Message();
                        deleteMs.setSender(id);
                        deleteMs.setList(deleteID);
                        deleteMs.setMesstype(5);
                        oos.writeObject(deleteMs);
                    } catch(IOException e1) {
                        e1.printStackTrace();
                    }
                }
            }
            @Override
            public void mouseClicked(MouseEvent e) {
                if(e.getSource() = = JL_prePage){
                    if(page ! = 1){
```

```java
                page--;
                for(int i = (page -1) * 5; i < (page -1) * 5 + 5; i++){
                    int k = i % 5;
                    if(i < nameList.size()){
                        label[k].setText(nameList.get(i).toString());
                        label[k].setForeground(new Color(187,187,187));
                        this.repaint();
                    }
                    Rectangle rs = label[k].getBounds();
                    if(rs.x != 30){
                        rs.x -= 130;
                        label[k].setBounds(rs);
                    }
                }
            }
            else{
                JOptionPane.showMessageDialog(null,"已经是第一页了");
            }
        }
        else if(e.getSource() == JL_nextPage) {
            if(page != maxPage){
                page++;
                for(int i = (page -1) * 5; i < (page -1) * 5 + 5; i++){
                    int k = i % 5;
                    if(i < nameList.size()) {
                        label[k].setText(nameList.get(i));
                        label[k].setForeground(new Color(187,187,187));
                        this.repaint();
                    }
                    else
                        label[k].setText("");
                    Rectangle rs = label[k].getBounds();
                    if(rs.x != 30){
                        rs.x -= 130;
                        label[k].setBounds(rs);
                    }
                }
            }
            else{
                JOptionPane.showMessageDialog(null,"已经是最后一页了");
            }
        }
```

```java
        else if(e.getClickCount() == 1){    //选中某条员工信息
            JLabel jlClick = (JLabel) e.getSource();
            if(jlClick.getForeground().equals(Color.blue)){
                jlClick.setForeground(new Color(187,187,187));
                Rectangle rs = jlClick.getBounds();
                rs.x -= 130;
                jlClick.setBounds(rs);
            }
            else{
                jlClick.setForeground(Color.blue);
                Rectangle rs = jlClick.getBounds();
                rs.x += 130;
                jlClick.setBounds(rs);
            }
        }
    }
    public void UpdateMaxPage(){
    maxPage = (nameList.size() % 5 == 0)? nameList.size() /5: nameList.size() /5 + 1;
    }
    @Override
    public void mouseEntered(MouseEvent e){
        //TODO Auto-generated method stub
    }
    @Override
    public void mouseExited(MouseEvent e){
        //TODO Auto-generated method stub
    }
    @Override
    public void mousePressed(MouseEvent e){
        //TODO Auto-generated method stub
    }
    @Override
    public void mouseReleased(MouseEvent e){
        //TODO Auto-generated method stub
    }
}
```

【代码7-38】修改 ClientSocketListener 类，增加对用户列表消息接收的处理

```java
public void run(){
    ...
        //后续与服务器端交互信息的代码加在这里；
    if(ms.getMesstype() == 2){
```

```
            String id = ms.getSender();
            String[] list = ms.getContent().split(" ");
            ManageStaffList.addList(id, list);
        }
            else if (ms.getMesstype() = = 3)
    ...
    }
```

(2) 服务器端：实现删除员工功能。首先增加客户服务进程 ClientHandler 对消息类型为 5 的消息对象的处理过程（参见代码 7 – 39），然后实现相应的数据库接口以删除指定的员工信息（参见代码 7 – 40）。

【代码 7 – 39】修改 ClientHandler 类，增加对删除消息的处理

```
public void run(){
...
    else if (ms.getMesstype() = = 2) {
        ObjectOutputStream oos = new ObjectOutputStream(
            ClientHandlerManager.getFunction ( ms.getSender ( )) .getS ( ) .getOutputStream());
        Message backMs = new Message();
        backMs.setMesstype(2);
        backMs.setSender(ms.getSender());
        backMs.setContent(new JDBC().applyStaffList());
        oos.writeObject(backMs);
    }
    else if(ms.getMesstype() = = 5) { //删除员工
        ArrayList list = ms.getList();
        new JDBC().deleteStaff(list);
    }
...
}
```

【代码 7 – 40】修改 JDBC 类，增加删除员工的 deleteStaff（）方法，其中 applyStaffList（）返回员工名单

```
public String applyStaffList(){//用来返回员工名单
    String staffNameList = null;
    try {
        Class.forName(driver);
        conn = DriverManager.getConnection(url, userName, sqlpassword);
        stmt = conn.createStatement();
        String sqlSelect = "SELECT id" + "FROM userInfo";
        rs = stmt.executeQuery(sqlSelect);
        while (rs.next()){
            if (rs.isFirst()){
```

```java
                staffNameList = rs.getString(1);
            }
            else {
                staffNameList += rs.getString(1);
            }
            if (rs.isLast() != true) {
                staffNameList += " ";
            }
        }
        rs.close();
        stmt.close();
        conn.close();
    } catch (ClassNotFoundException e) {
        e.printStackTrace();
    } catch (SQLException e) {
        e.printStackTrace();
    }

    return staffNameList;
}
public void deleteStaff(ArrayList list){ //删除员工
    System.out.println("运行");
    try {
    Class.forName(driver);
    conn = DriverManager.getConnection(url, userName, sqlpassword);
    stmt = conn.createStatement();
    for (int i = 0; i < list.size(); i++){
        String sql = "delete from userInfo" + "where id ='" + list.get(i) + "'";
        stmt.executeUpdate(sql);
        }
    } catch (ClassNotFoundException e) {
        e.printStackTrace();
    } catch (SQLException e) {
        e.printStackTrace();
    }
  }
}
```

5. 查看员工信息功能

本部分实现如图 7-17 所示的员工信息查看界面。其中界面的左侧是用户查询和用户名单，员工名单由 JList 技术实现；界面右侧是选中的员工信息。查看员工信息的步骤如下：

（1）首先在查询界面向服务器查询所有员工名单，并将名单显示在界面左侧的名单列表中。

（2）经理用户在查询栏填写用户名关键字，员工信息查询界面通过监听查询栏中的用户输入，从员工名单中查询出只含有用户名关键字的用户名列表。

（3）经理用户双击【员工列表】中的员工名（ID），被选中的用户名被更新在查询栏内。

（4）经理用户单击【查看】按钮，界面就会向服务器请求显示选中用户的所有信息（不包括密码）；等服务器接受请求后，所选中的员工信息即可显示在界面的右侧。如图 7 – 36 所示。

图 7 – 36　查看员工详细信息界面

实现查看员工详细信息功能包含以下几个步骤：

（1）客户端：实现在客户端功能选择界面的【查看员工信息】按钮的事件处理。

【代码 7 – 41】修改 ClientMainUI 类，增加【查看员工信息】按钮的事件处理

```
public void actionPerformed(ActionEvent e){
…
    else if (e.getSource() = = Deleteemployee){
        if(judgeAuthority(authority)){
            new DeleteStaffUI(id);
        }
    }
…
}
```

(2) 客户端：实现查看员工信息界面和事件处理。

【代码 7-42】 CheckStaffInformation 类

```java
package staffMessFunction;
import java.awt.Color;
import java.awt.event.*;
import java.io.IOException;
import java.io.ObjectOutputStream;
import java.util.ArrayList;
import javax.swing.*;
import javax.swing.event.DocumentEvent;
import ManageClientConnServer.*;
import common.Message;
public class CheckStaffInformation extends JFrame{
    private JButton find;
    private JTextField nameField;
    private String id;
    protected JTextField JT_id = new JTextField("");
    protected JTextField JT_authority = new JTextField("");
    protected JTextField JT_income = new JTextField("");
    protected JTextField JT_age = new JTextField("");
    protected JTextField JT_sex = new JTextField("");
    protected JTextField JT_Address = new JTextField("");
    protected JTextField JT_telNum = new JTextField("");
    protected JTextField JT_QQnum = new JTextField("");
    protected JTextField[] jts = { JT_id, JT_authority, JT_income, JT_age, JT_sex, JT_Address, JT_telNum, JT_QQnum };
    public CheckStaffInformation(String id){
        this.id = id;
        this.setSize(600,480);
        this.setLocation(300,120);
        try {
            Message ms = new Message();
            ms.setMesstype(2);
            ms.setSender(id);
            ms.setGetter(id);
            ObjectOutputStream oos = new ObjectOutputStream( ClientSocket.getClientConnServer(id).getS().getOutputStream());
            oos.writeObject(ms);
            Thread.sleep(100);
        }catch(IOException e){
            e.printStackTrace();
```

```java
        }catch(InterruptedException e){
            e.printStackTrace();
        }
        String[] list = ManageStaffList.getStringList(id);
        this.add(new GUI(list));
        this.setVisible(true);
    }
    //内部类 GUI 实现了查看员工信息界面及员工列表的事件处理
    class GUI extends JPanel implements MouseListener, ActionListener{
        private JList lst;
        private String[] str;
        final DefaultListModel iItems = new DefaultListModel();
        public GUI(final String[] str){
            this.str = str;
            this.setLayout(null);
            this.setBackground(new Color(1,100,124));
            this.setSize(520,500);
            this.setLocation(450,150);
            JT_id.setBounds(380,30,150,30);
            JT_authority.setBounds(380,80,150,30);
            JT_income.setBounds(380,130,150,30);
            JT_age.setBounds(380,180,150,30);
            JT_telNum.setBounds(380,230,150,30);
            JT_Address.setBounds(380,280,150,30);
            JT_sex.setBounds(380,330,150,30);
            JT_QQnum.setBounds(380,380,150,30);
            JLabel[] jt_labels =
            { new JLabel("姓名:"), new JLabel("权限:"),
              new JLabel("收入:"), new JLabel("年龄:"),
              new JLabel("性别:"), new JLabel("联系电话:"),
              new JLabel("家庭住址:"), new JLabel("QQ号码:")
            };
            jt_labels[0].setBounds(300,30,150,30);
            jt_labels[1].setBounds(300,80,150,30);
            jt_labels[2].setBounds(300,130,150,30);
            jt_labels[3].setBounds(300,180,150,30);
            jt_labels[4].setBounds(300,330,150,30);
            jt_labels[5].setBounds(300,280,150,30);
            jt_labels[6].setBounds(300,230,150,30);
            jt_labels[7].setBounds(300,380,150,30);
            for(int i = 0; i < jts.length; i++){
                jts[i].setEditable(false);
```

```java
            jt_labels[i].setForeground(new Color(187,187,187));
            this.add(jt_labels[i]);
            this.add(jts[i]);
        }
        nameField = new JTextField();
        nameField.setBounds(20,40,142,25);
        nameField.getDocument().addDocumentListener(new javax.swing.event.DocumentListener(){
            public void changedUpdate(DocumentEvent e) {
                //Do nothing
            }
            @Override
            public void insertUpdate(DocumentEvent e) {
                String s = nameField.getText().trim();
                iltems.removeAllElements();
                for(int i = 0; i < str.length; i++){
                    if(str[i].indexOf(s) > -1){
                        iltems.addElement(str[i]);
                    }
                }
                lst.setModel(iltems);
            }
            @Override
            public void removeUpdate(DocumentEvent e) {
                if(nameField.getText().equals("")){
                    iltems.removeAllElements();
                    lst.setModel(iltems); //DefaultListModel
                    for(int i = 0; i < str.length; i++) {
                        iltems.addElement(str[i]);
                    }
                }
                else{
                    iltems.removeAllElements();
                    String s = nameField.getText().trim();
                    lst.setModel(iltems); //DefaultListModel
                    for(int i = 0; i < str.length; i++){
                        if(str[i].indexOf(s) > -1) {
                            iltems.addElement(str[i]);
                        }
                    }
                }
            }
```

```java
            });
            this.add(nameField);
            ImageIcon entered = new ImageIcon("src/image/staffManage/查看2.png");
            ImageIcon released = new ImageIcon("src/image/staffManage/查看3.png");
            ImageIcon pressed = new ImageIcon("src/image/staffManage/查看1.png");
            find = new JButton();
            find.setOpaque(false);//透明化
            find.setBorderPainted(false);
            find.setBorder(null);
            find.setBounds(180,40,92,33); //按钮大小,位置
            find.setContentAreaFilled(false);//按钮透明
            find.setPressedIcon(pressed);//按下按钮时的图片
            find.setRolloverIcon(entered);//经过时的图片
            find.setIcon(released);//初始化
            find.addActionListener(this);
            this.add(find);
            lst = new JList();
            lst.setModel(iltems);
            for (int i = 0; i < str.length; i++){
                iltems.addElement(str[i]);
            }
            JScrollPane scroll = new JScrollPane(lst);
            scroll.setBounds(20,90,225,330);
            lst.addMouseListener(this);
            this.add(scroll);
            this.setVisible(true);
        }
        @Override
        public void actionPerformed(ActionEvent e) {
            if (e.getSource() == find) {
            try {
                ObjectOutputStream oos = new ObjectOutputStream (ClientSock-
et.getClientConnServer(id).getS().getOutputStream());
                Message ms = new Message();
                ms.setMesstype(3);
                ms.setSender(nameField.getText());
                ms.setGetter(id);
                oos.writeObject(ms);
                Thread.sleep(200);
                }catch (IOException e1) {
                e1.printStackTrace();
            }catch (InterruptedException e1) {
```

```java
            e1.printStackTrace();
        }
        ArrayList msList = ManageMsPerson.getArrayList(id);
        for (int i = 0; i < msList.size(); i++) {
            JT_id.setText(msList.get(0) + "");
            JT_authority.setText(msList.get(1) + "");
            JT_income.setText(msList.get(3) + "");
            JT_age.setText(msList.get(4) + "");
            JT_sex.setText(msList.get(5) + "");
            JT_Address.setText(msList.get(6) + "");
            JT_telNum.setText(msList.get(7) + "");
            JT_QQnum.setText(msList.get(8) + "");
        }
    }
}
@Override
public void mouseClicked(MouseEvent e) {
    int index = lst.locationToIndex(e.getPoint());
    nameField.setText((String) iltems.get(index));
}
@Override
public void mouseEntered(MouseEvent e) {
    //TODO Auto-generated method stub
}
@Override
public void mouseExited(MouseEvent e) {
    //TODO Auto-generated method stub
}
@Override
public void mousePressed(MouseEvent e) {
    //TODO Auto-generated method stub
}
@Override
public void mouseReleased(MouseEvent e) {
    //TODO Auto-generated method stub
    }
  }
}
```

7.3.3 增量7-3：实现公告

增量7-3旨在实现办公管理系统的公告功能。在公告功能中，我们需要实现以下几个界面：

（1）实现如图7-25所示的查看公告首页。界面要支持分页显示，每页显示多条（如8条）公告，通过界面上的【上一页】和【下一页】按钮可以浏览更多的公告。每条公告的信息包含公告标题、发布日期和时间。当鼠标移到公告标题时，鼠标会变成手状，双击鼠标即可查看公告的详细内容。同时，在公告首页有【删除】按钮，具有经理权限的用户可以删除选中的公告。

（2）实现如图7-26所示的查看公告详细信息界面，界面包含公告标题和公告内容。

（3）实现对图7-25选中的公告进行删除，具有经理权限的用户可以删除公告，在删除前系统会提醒用户是否"确认删除"，以避免用户的误操作。

（4）实现如图7-28所示的发布公告界面，具有经理权限的用户可以发布新的公告，通过客户端功能选择界面的【发布公告】按钮实现。

1. 公告功能：查看和删除公告

查看和删除公告的功能可通过以下几个步骤实现：

（1）客户端：实现功能选择界面对【查看公告】和【删除公告】按钮的事件处理。

【代码7-43】修改 ClientMainUI 类，增加公告功能事件处理

```
public void actionPerformed(ActionEvent e){
…
//公告的按钮
    if(e.getSource() = = Checkannouncement){
        new PublicationBrowser (id, authority);
    }
    else if(e.getSource() = = Announcement){
        if(judgeAuthority(authority)){
            new PublicationSend(id);
        }
    }
…
}
```

（2）客户端：实现公告首页、查看公告详细信息和删除公告。

PublicationBrowser 类（参见代码7-44）实现了查看公告首页、查看公告详细信息和删除公告界面。其中，构造方法 PublicationBrowser（String id, String authority）首先通过消息（消息类型为11）从服务器获取公告列表，如果暂时没有公告，则弹出【无公告】对话框，否则调用方法 GUI（）显示公告首页。PublicationBrowser 类中的 GUI2（）方法实现了查看详细公告界面效果，方法 GUI3（）实现了删除公告界面效果，nextPage（）和 PPage（）方法分别实现了【下一页】和【上一页】的翻页功能。

PublicationBrowser 类通过扩展 MouseListener 接口实现了界面对鼠标操作的响应。当鼠标移动到【公告标题】、【上一页】和【下一页】时，mouseEntered（）方法即可捕

获到此鼠标事件,同时鼠标会变成手状,这时单击鼠标即可查看公告详细内容,或者实现界面跳转;如果用户单击鼠标左键,mousePressed()方法就可捕获到该事件,实现查看公告详细内容,或者界面上下页跳转的效果。

PublicationBrowser 类通过扩展 ActionListener 接口实现了对事件的处理,当用户单击【删除】按钮时,事件的处理将由 actionPerformed()方法执行。如果是经理用户,则弹出【确认删除】对话框,否则弹出【用户权限不足】对话框;如果用户选择【确认继续】,PublicationBrowser 将会调用 GUI3()方法提示用户是否确认删除,如果用户确定删除公告,GUI3()方法将向服务器发送一个公告删除消息(消息类型为 12),服务器收到后就会删除此公告,然后 GUI3()方法会更新查看公告界面显示的内容。

PublicationBrowser 类通过扩展 ItemListener 接口实现了对复选框 CheckBox 的事件处理,用于统计用户选择了哪条公告。在用户单击【删除】按钮时,GUI3()方法会根据用户的选择删除相应的公告信息。

【代码 7-44】PublicationBrowser 类

```
package publication;
import java.awt.Checkbox;
import java.awt.Cursor;
import java.awt.Font;
import java.awt.event.*;
import java.io.IOException;
import java.io.ObjectOutputStream;
import java.util.ArrayList;
import javax.swing.*;
import ManageClientConnServer.ClientSocket;
import ManageClientConnServer.ManagePublic;
import common.Message;
public class PublicationBrowser extends JFrame implements ActionListener,
MouseListener, ItemListener{
    private String id;
    private ArrayList publication;
    private int length;
    private Boolean num[];
    private JLabel label[];
    private JLabel labelDate[];
    private JButton buttonDelete;
    private int index;
    private JLabel labelT;
    private JLabel pageNext;
    private JLabel pageP;
    private JTextArea textM;
    private Cursor cursor;
```

```java
private Checkbox check[];
private int pages = 1;//页数标志
private final int pageNum = 8;//每页八个公告
private int number = 0;//从零开始
private ImageIcon image;
private JLabel labelImage;
private int flag = 0;//0 时代表没有公告被选中,点删除不作出反应
private JPanel root;
private String authority;
public PublicationBrowser(String id, String authority){
    this.id = id;
    this.authority = authority;
    publication = new ArrayList();
    Message ms = new Message();
    ms.setMesstype(11);
    ms.setSender(id);
    ms.setGetter(id);
    try{
            ObjectOutputStream oos = new ObjectOutputStream(ClientSocket.getClientConnServer(id).getS().getOutputStream());
        oos.writeObject(ms);
        Thread.sleep(200);
    }catch (IOException e1) {
        e1.printStackTrace();
    }catch (InterruptedException e) {
        e.printStackTrace();
    }
    publication = ManagePublic.getArrayList(id);
    if (publication ! = null){
        length = publication.size();
        if (length < = 0){
            JOptionPane.showMessageDialog(null,"暂时没有公告");
        }
        else{
            GUI();
        }
    }
}
    private void GUI(){//公告首页
        this.setSize(662,513);
        this.setTitle("公告");
        this.setLocationRelativeTo(null);
```

```java
this.setLayout(null);
image = new ImageIcon("src/image/publicTell/公告背景.png");
labelImage = new JLabel(image);
this.getLayeredPane().add(labelImage, new Integer(Integer.MIN_VALUE));
        labelImage.setBounds(0, 0, image.getIconWidth(), image.getIconHeight());
root = (JPanel) this.getContentPane();
root.setOpaque(false);
num = new Boolean[length /3];
for(int i = 0; i < length /3; i++) {
  num[i] = false;
}
label = new JLabel[length /3];
labelDate = new JLabel[length /3];
check = new Checkbox[length /3];
//按钮美化
   ImageIcon entered = new ImageIcon("src/image/publicTell/删除xz.png");
   ImageIcon released = new ImageIcon("src/image/publicTell/删除yx.png");
   ImageIcon pressed = new ImageIcon("src/image/publicTell/删除ax.png");
buttonDelete = new JButton();
buttonDelete.setOpaque(false);//透明化
buttonDelete.setIcon(released);//初始化
buttonDelete.setBorderPainted(false);
buttonDelete.setBorder(null);
buttonDelete.setBounds(190,350,90,32);//按钮大小,位置
buttonDelete.setContentAreaFilled(false);//按钮透明
buttonDelete.setPressedIcon(pressed);//按下按钮时的图片
buttonDelete.setRolloverIcon(entered); //经过时的图片
buttonDelete.addActionListener(this);
pageNext = new JLabel("下一页");
pageNext.setBounds(370,350,40,26);
pageNext.addMouseListener(this);
pageP = new JLabel("上一页");
pageP.setBounds(300,350,40,26);
pageP.addMouseListener(this);
for(int i = 0, j = 0, m = 2, n = 80; j < length /3 && j < pages * pageNum;
    i += 3, j++, m += 3, n += 30){
```

```java
            label[j] = new JLabel(publication.get(i) + "");
            label[j].setSize(200, 40);
            label[j].setLocation(150, n);
            label[j].setHorizontalAlignment(SwingConstants.LEFT);
            label[j].addMouseListener(this);
            labelDate[j] = new JLabel((publication.get(m) + "").substring(0, (publication.get(m) + "").indexOf(":") + 3));
            labelDate[j].setBounds(420, n, 150, 30);
            check[j] = new Checkbox();
            check[j].setLocation(110, n + 14);
            check[j].setSize(12, 12);
            check[j].addItemListener(this);
            this.add(label[j]);
            this.add(labelDate[j]);
            this.add(check[j]);
            this.add(buttonDelete);
        }
        cursor = new Cursor(HAND_CURSOR);
        this.add(buttonDelete);
        this.add(pageNext);
        this.add(pageP);
        this.setVisible(true);
    }
    @Override
    public void actionPerformed(ActionEvent e){
        if(authority.equals("mgr")){
            if(e.getSource() == buttonDelete){
                if(flag == 1){
                    GUI3();
                }
                else{
                    JOptionPane.showMessageDialog(null, "没选中公告");
                }
            }
        }
        else{
            JOptionPane.showMessageDialog(null, "权限不足");
        }
    }
    public void GUI2(){//查看详细公告窗口
        JFrame frameBrowser = new JFrame("查看公告");
        ImageIcon bg = new ImageIcon("src/image/publicTell/公告背景.png");
```

```java
            frameBrowser = new JFrame("主界面");
            frameBrowser.setSize(662, 513);
            frameBrowser.setLocation(300, 200);
            frameBrowser.setLayout(null);
             frameBrowser.getLayeredPane().add(label, new Integer(Integer.MIN_VALUE));
            JLabel label = new JLabel(bg);
            label.setBounds(0, 0, bg.getIconWidth(), bg.getIconHeight());
            JPanel jp = (JPanel) frameBrowser.getContentPane();
            jp.setOpaque(false);
            labelT = new JLabel(publication.get(3 * index) + "");
            labelT.setFont(new Font("黑体", 1, 20));
            labelT.setBounds(getWidth() / 2 - labelT.getWidth() / 2 - 37, 80, 100, 40);
            frameBrowser.add(labelT);
            textM = new JTextArea(20, 10);
            textM.setLineWrap(true);
            textM.setEditable(false);
            textM.setText(publication.get(3 * index + 1) + "");
            textM.setEditable(false);
            textM.setBounds(200, 150, 300, 200);
            textM.setOpaque(false);
            frameBrowser.add(textM);
            frameBrowser.setVisible(true);
        }
    public void GUI3(){//删除公告确认窗口
        int select = JOptionPane.showConfirmDialog(null, "是否删除?", "确认窗口", JOptionPane.YES_NO_OPTION);
            if(select == JOptionPane.YES_OPTION){
            ArrayList list = new ArrayList();
            Message ms = new Message();
            ms.setMesstype(12);
            ms.setSender(id);
            ms.setGetter(id);
            for(int i = 0; i < check.length; i++){
                if(num[i] == true) {
                    list.add(publication.get(i * 3));
                    this.remove(label[i]);
                    this.remove(check[i]);
                    this.remove(labelDate[i]);
                }
            }
```

```java
            ms.setList(list);
            try{
                    ObjectOutputStream oos = new ObjectOutputStream(ClientSock-
et.getClientConnServer(id).getS().getOutputStream());
                    oos.writeObject(ms);
                    this.repaint();
                    Thread.sleep(100);
            }catch (IOException e1) {
                    e1.printStackTrace();
            }catch (InterruptedException e) {
                    e.printStackTrace();
            }
            JOptionPane.showMessageDialog(null,"删除成功");
            validate();
        }
    }


    @Override
    public void mouseEntered(MouseEvent e) {
        for(int i = 0; i < length /3; i++){
            if(e.getSource() == label[i]){
                label[i].setCursor(cursor);
            }
}
            if(e.getSource() == pageNext) {
                pageNext.setCursor(cursor);
            }
            else if(e.getSource() == pageP) {
                pageP.setCursor(cursor);
            }
    }
    @Override
    public void mousePressed(MouseEvent e) {
            for(int i = 0; i < length /3; i++) {
                if(e.getSource() == label[i]){
                    index = i;
                        GUI2();
                }
            }
                if(e.getSource() == pageNext) {
                    nextPage();
                }
```

```java
            else if (e.getSource() = = pageP) {
                PPage();
            }
        }
    }
    private void nextPage() {
        if (sum() % 8 ! = 0) {
        JOptionPane.showMessageDialog(null, "已经是最后一页");
        }
        else if (length /3 < = pageNum * pages) {
            JOptionPane.showMessageDialog(null, "已经是最后一页");
        }
        else {
            pages + +;
            number + = 8;
            root.removeAll();
            repaint();
            updateNext();
        }
    }
    private int sum() {
        int sum = 0;
        for (int i = 0; i < num.length; i + +){
            if (num[i] = = true)
                    + +sum;
        }
        return sum;
    }
    private void updateP() {
        System.out.println(pages * pageNum);
        for (int i = 3 * number, k = number, m = number * 3 + 2, n = 80; k < pages * pageNum && k < length /3; i + = 3, m + = 3, k + +, n + = 30)
        {
        if (num[k] = = true) continue;
        label[k] = new JLabel(publication.get(i) + "");
        label[k].setHorizontalAlignment(SwingConstants.LEFT);
        label[k].setSize(200, 40);
        label[k].setLocation(150, n);
        label[k].addMouseListener(this);
        labelDate[k] = new JLabel((publication.get(m) + "").substring(0, (publication.get(m) + "").indexOf(":") + 3));
        labelDate[k].setBounds(420, n, 150, 30);
        check[k] = new Checkbox();
```

```java
                check[k].setLocation(110, n + 14);
                check[k].setSize(12, 12);
                check[k].addItemListener(this);
                this.add(label[k]);
                this.add(labelDate[k]);
                this.add(check[k]);
            }
        this.add(buttonDelete);
        this.add(pageNext);
        this.add(pageP);
        validate();
    }
    private void updateNext(){
        for (int i = 3 * number, k = number, m = number * 3 + 2, n = 80; k < length /3 && k < pages * pageNum; i + = 3, m + = 3, k + +, n + = 30)
        {
                if (num[k] = = true)
                continue;
                label[k] = new JLabel(publication.get(i) + "");
                label[k].setHorizontalAlignment(SwingConstants.LEFT);
                label[k].setSize(200, 40);
                label[k].setLocation(150, n);
                label[k].addMouseListener(this);
                labelDate[k] = new JLabel((publication.get(m) + "").substring(0, (publication.get(m) + "").indexOf(":") + 3));
                labelDate[k].setBounds(420, n, 150, 30);
                check[k] = new Checkbox();
                check[k].addItemListener(this);
                check[k].setLocation(110, n + 14);
                check[k].setSize(12, 12);
                this.add(label[k]);
                this.add(labelDate[k]);
                this.add(check[k]);
            }
        this.add(buttonDelete);
        this.add(pageNext);
        this.add(pageP);
        validate();
    }
    @ Override
    public void itemStateChanged(ItemEvent e) {
        int stateChange;
```

```java
            Object source = e.getItemSelectable();
            for(int i = 0; i < check.length; i++){
                if(source == check[i]){
                    stateChange = e.getStateChange();
                    if(stateChange == ItemEvent.SELECTED){
                        flag = 1;
                        num[i] = true;
                    }
                    else if(stateChange == ItemEvent.DESELECTED){
                        flag = 0;
                        num[i] = false;
                    }
                }
            }
        }
        @Override
        public void mouseReleased(MouseEvent e) {
            //TODO Auto-generated method stub
        }
        @Override
        public void mouseClicked(MouseEvent e) {
            //TODO Auto-generated method stub
        }
        @Override
        public void mouseExited(MouseEvent e) {
            //TODO Auto-generated method stub
        }
    }
```

（3）客户端：实现客户端Socket监听器对接收到的公告列表的处理。

客户端Socket监听器线程接收到类型为11的查看公告列表消息后，从中取出消息列表（参见代码7-45），存入MessagePublicationList（参见代码7-46）中，这样查看公告界面就可以从中读取公告信息。

【代码7-45】修改ClientSocketListener类，增加处理公告列表代码

```java
public void run() {
    ...
    //后续与服务器端交互信息的代码加在这里；
    else if(ms.getMesstype() == 11) //查看公告类型为11
    {
        String id = ms.getSender();
        ArrayList list = ms.getList();
        MessagePublicationList.addArrayList(id, list);
```

        }
    ...
    }

【代码 7 – 46】表 ManagePublicationList 类
```
package ManageClientConnServer;
import java.util.ArrayList;
import java.util.HashMap;
public class ManagePublicationList {
    private static HashMap Public_Hash = new HashMap < String , ArrayList > ();
    public static void addArrayList(String id , ArrayList msPersonList ) {
        Public_Hash.put( id, msPersonList);
    }
    public static ArrayList getArrayList(String id) {
        return (ArrayList)Public_Hash.get(id);
    }
    public static void addArray(String id,int day[ ]) {
        Public_Hash.put(id, day);
    }
    public static int[ ] getArray(String id){
        return (int[ ])Public_Hash.get(id);
    }
}
```

　　(4) 服务器端：对查看/删除公告的消息处理。客户服务线程 ClientHandler 在接收到类型为 11 的 Message 对象（参见代码 7 – 47）时，首先调用 JDBC 类的 getPublicationList () 方法（需要在 JDBC 类中增加，参见代码 7 – 48），从数据库的公告表中读取公告信息列表，然后把此列表放在类型为 11 的 Message 对象中，反馈给客户端，客户端的 Socket 监听器线程负责接收该消息列表，并存入 MessagePublicationList 中。

　　客户服务线程 ClientHandler 在接收到类型为 12 的 Message 对象（参见代码 7 – 47）时，调用 JDBC 类的 deletePublicationItem () 方法（需要在 JDBC 类中增加，参见代码 7 – 48），把指定的公告从数据库中删除。

【代码 7 – 47】修改 ClientHandler 类，增加对类型为 11、12 的 Message 的处理
```
public void run() {
    ...
        else if (ms.getMesstype() = = 11) { //接受查看公告申请
            String id = ms.getSender();
            ArrayList publication = new JDBC().getPublicationList();
            ObjectOutputStream oos = new ObjectOutputStream(s.getOutputStream());
            Message publicMs = new Message();
            publicMs.setSender(id);
```

```java
            publicMs.setMesstype(11);
            publicMs.setList(publication);
            oos.writeObject(publicMs);
        }
        else if (ms.getMesstype() == 12) { //接受删除公告的申请
            ArrayList list = ms.getList();
            new JDBC().deletePublicationItem(list);
        }
    ...
    }
```

【代码7-48】修改JDBC类,增加实现查看公告列表getPublicationList( )方法和删除公告deletePublicationItem( )方法

```java
public ArrayList getPublicationList(){//申请查看公告列表
    ArrayList publication = new ArrayList();
    try {
        Class.forName(driver);
        conn = DriverManager.getConnection(url, userName, sqlpassword);
        stmt = conn.createStatement();
        String sql = "select title,message,time from publication order by time desc";
        rs = stmt.executeQuery(sql);
        while (rs.next()) {
            publication.add(rs.getString(1));
            publication.add(rs.getString(2));
            publication.add(rs.getTimestamp(3));
        }
        rs.close();
        stmt.close();
        conn.close();
    } catch (ClassNotFoundException e) {
        e.printStackTrace();
    } catch (SQLException e) {
        e.printStackTrace();
    }

    return publication;
}

public void deletePublicationItem(ArrayList list) { //删除公告
    try {
        Class.forName(driver);
```

```
            conn = DriverManager.getConnection(url, userName, sqlpassword);
            PreparedStatement pstmt = null;
            String sql = "delete from publication where title = ?";
            pstmt = conn.prepareStatement(sql);
            for (int i = 0; i < list.size(); i + +) {
                pstmt.setString(1, list.get(i).toString());
                pstmt.addBatch();
            }
            int trmp[] = pstmt.executeBatch();
            pstmt.close();
            conn.close();
        }catch (ClassNotFoundException e) {
            e.printStackTrace();
        }catch (SQLException e) {
            e.printStackTrace();
        }
    }
```

2. 发布公告功能

经理用户单击【发布公告】按钮后可弹出发布公告界面，如图 7 - 28 所示。单击界面上的【发布公告】按钮，客户端就会发送一个发布公告消息（消息类型为 10）对象给服务器。服务器收到此消息后，把公告信息插入数据库的 publication 数据表中，以实现成功公告发布功能。其他用户不可以发布公告。发布公告具体的实现步骤如下：

（1）客户端：实现发布公告。首先，在功能选择界面 ClientMainUI 类中，增加对【发布公告】按钮的事件处理（代码 7 - 49）。然后实现发布公告界面以及【发布公告】按钮的事件处理（代码 7 - 50），以实现发送发布公告消息给服务器（代码 7 - 51）。

【代码 7 - 49】修改 ClientMainUI 类，增加公告功能事件处理

```
public void actionPerformed(ActionEvent e) {
    …
        //增加公告的按钮
        else if (e.getSource() = = Announcement) {
          if (judgeAuthority(authority)){
            new PublicationSend(id);
            }
        }
    …
}
```

【代码 7 - 50】NewPublication 类

```
public class PublicationSend extends JFrame implements ActionListener {
    private JLabel labelT, labelM;
    private JTextField textT;
    private JTextArea textM;
```

```java
    private JButton buttonSend;
    private String title, message, id;
    private JFrame frame;
    public publicSend(String id) {
        this.id = id;
        frame = new JFrame("发布公告");
        frame.setLocationRelativeTo(null);
        frame.setSize(662,513);
        frame.setLocation(300,200);
        frame.setLayout(null);
        ImageIcon bg = new ImageIcon("src/image/publicTell/公告背景.png");
        JLabel label = newJLabel(bg);
        jp.setOpaque(false);
        labelT = new JLabel("标题");
        labelT.setBounds(160,80,100,40);
        labelM = new JLabel("正文");
        labelM.setBounds(100,220,50,50);
        textT = new JTextField(10);
        textT.setBounds(190,86,250,25);
        textT.setOpaque(false);
        textM = new JTextArea(20,20);
        textM.setLineWrap(true);
        textM.setBounds(170,150,300,200);
        ImageIcon entered = new ImageIcon("src/image/publicTell/发布1.png");
        ImageIcon released = new ImageIcon("src/image/publicTell/发布2.png");
        ImageIcon pressed = new ImageIcon("src/image/publicTell/发布3.png");
        buttonSend = new JButton();
        buttonSend.setOpaque(false);           //透明化
        buttonSend.setIcon(released);          //初始化
        buttonSend.setBorderPainted(false);
        buttonSend.setBorder(null);
        setBounds(190,170,117,37);             //按钮大小,位置
        buttonSend.setContentAreaFilled(false);//按钮透明
        buttonSend.setPressedIcon(pressed);
        buttonSend.setRolloverIcon(entered);   //经过时的图片
        buttonSend.setBounds(270,380,90,30);
        buttonSend.addActionListener(this);
        frame.add(labelT);
        frame.add(labelM);
```

```
            frame.add(textT);
            frame.add(textM);
            frame.add(buttonSend);
            frame.setVisible(true);
    }
    @Override
    public void actionPerformed(ActionEvent e) {
        title = textT.getText();
        message = textM.getText();
        new SendNewPublicationToServer(id, title, message, frame);
        JOptionPane.showMessageDialog(null, "发布成功");
    }
}
```

【代码7-51】 SendNewPublicationToServer 类

```
package publication;
import java.io.IOException;
import java.io.ObjectOutputStream;
import javax.swing.JFrame;
import ManageClientConnServer.ClientSocket;
import common.Message;
public class SendNewPublicationToServer{
    private String title;
    private String message;
    private String id;
    private static JFrame frame;
    public SendNewPublicationToServer(String id, String title, String message, JFrame frame) {
        this.id = id;
        this.title = title;
        this.message = message;
        this.frame = frame;
        Message ms = new Message();
        ms.setContent(title + "Fl_Ag" + message);
        ms.setMesstype(10);
        try{
            ObjectOutputStream oos = new ObjectOutputStream(ClientSocket.getClientConnServer(id).getS().getOutputStream());
            oos.writeObject(ms);
        } catch (IOException e1) {
            e1.printStackTrace();
        }
    }
}
```

（2）服务器端：处理发布公告消息。客户服务线程 ClientHandler 在接收到类型为 10 的 Message 对象（参见代码 7-52）时，调用 JDBC 类的 addPublicationItem（）方法（需要在 JDBC 类中增加，参见代码 7-53），addPublicationItem（）方法向数据库的公告表中插入一个新的公告记录，即可完成公告发布过程。

【代码 7-52】修改 ClientHandler 类，增加对类型为 10 的发布公告 Message 的处理

```
public void run(){
...
    else if(ms.getMesstype() == 10){  //发布公告
        String content = ms.getContent();
        String title = content.substring(0,content.indexOf("Fl_Ag"));
        String message = content.substring(content.indexOf("Fl_Ag") + 5,content.length());
        String returnValue = new JDBC().addPublicationItem(title,message);
    }
...
}
```

【代码 7-53】修改 JDBC 类，增加插入公告的 addPublicationItem（）方法

```
//添加公告
public String addPublicationItem(String title,String message){
    try{
        Class.forName(driver);
        conn = DriverManager.getConnection(url,userName,sqlpassword);
        stmt = conn.createStatement();
        String sql = "insert into publication (title,message) " + " values ('"
            + title + "','" + message + "')";
        stmt.executeUpdate(sql);
        return "true";
    }catch(ClassNotFoundException e){
        e.printStackTrace();
    }catch(SQLException e){
        e.printStackTrace();
    }
    return "false";
}
```

7.3.4 增量 7-4：实现考勤

增量 7-4 旨在实现办公管理系统的考勤功能。

1. 考勤签到

用户通过单击客户端功能选择界面中的【签到】按钮即可实现当天签到。如果用户当天已经签到，则弹出【今天已签到】对话框（如图 7-21 所示），否则发送签到消

息（消息类型为 50）给服务器，并显示【签到成功】对话框（如图 7 – 18 所示）。

下面我们分步依次实现考勤功能：

(1) 客户端：实现功能选择界面对【签到】按钮的事件处理。

【代码 7 – 54】修改 ClientMainUI 类，增加签到事件处理

```java
public void actionPerformed(ActionEvent e){
    ...
    else if(e.getSource() == Signin){
        if(flag == 0){//用户还没有签到
            flag = 1;
            new RegisterAttendance(id).writeIn();
            JOptionPane.showMessageDialog(null,"签到成功");
        }
        else if(flag == 1){
            JOptionPane.showMessageDialog(null,"你今天已经签到");
        }
    }
    ...
}
```

(2) 客户端：实现签到功能的 RegisterAttendance 类。

【代码 7 – 55】RegisterAttendance 类

```java
package attendance;
import java.io.IOException;
import java.io.ObjectOutputStream;
import javax.swing.JFrame;
import ManageClientConnServer.ClientSocket;
import common.Message;
public class RegisterAttendance extends JFrame{
    private String id;
    public RegisterAttendance(String id){
        this.id = id;
    }
    public void writeIn(){ //设置签到的消息类型为50
        Message ms = new Message();
        ms.setMesstype(50);
        ms.setSender(id);
        ms.setGetter(id);
        try{
            ObjectOutputStream oos = new ObjectOutputStream(new ClientSocket().getClientConnServer(id).getS().getOutputStream());
            oos.writeObject(ms);
            Thread.sleep(100);
```

```
        }catch(IOException e1){
            e1.printStackTrace();
        }catch(InterruptedException e){
            e.printStackTrace();
        }
    }
}
```

(3) 服务器端：实现签到功能。

首先增加客户服务线程对签到消息的处理（代码7-56），然后实现数据库插入签到记录的 JDBC 类的接口 addAttendance（代码7-57）。

【代码7-56】修改 ClientHandler 类，增加插入签到记录处理

```
public void run(){
...
    //用来处理签到
    else if(ms.getMesstype()==50){
        String id = ms.getSender();
        String content = new JDBC().addAttendance(id);
    }
...
}
```

【代码7-57】修改 JDBC 类，增加签到功能处理方法

```
public String addAttendance(String id){//用来处理签到功能
    String res = "Yes";
    try{
        Class.forName(driver);
        conn = DriverManager.getConnection(url, userName, sqlpassword);
        System.out.println("数据库连接成功");
        stmt = conn.createStatement();
        Date time = new Date();
        String sql = "insert into attend(id,present) values('" + id + "','" + res + "')";
        stmt.executeUpdate(sql);
        stmt.close();
        conn.close();
    }catch(ClassNotFoundException e){
        e.printStackTrace();
    }catch(SQLException e){
        e.printStackTrace();
    }
    return res;
}
```

2. 查看考勤

经理用户可以查看员工的出勤记录，在查看考勤记录界面（如图7-37所示）上，用户可以通过下拉菜单选择员工 ID 和考勤记录年份，根据选定的员工 ID 和年份，系统会自动从服务器读取相应的考勤记录，并把结果显示在界面上（如图7-38所示的记录）。

图7-37 查看考勤界面

图7-38 考勤记录

下面我们分步依次实现考勤功能：

（1）客户端：实现功能选择界面对【查看考勤】按钮的事件处理。

【代码7-58】修改 ClientMainUI 类，增加【查看考勤】按钮事件处理

```
public void actionPerformed(ActionEvent e) {
...
    else if (e.getSource() = = Checkattendance){
    if (judgeAuthority(authority)){
        new RequestAttendanceData(id);
    }
  }
...
}
```

（2）客户端：实现"查看考勤"功能的 RequestAttendanceData 类和 CheckAttendance 类。RequestAttendaceData 类实现向服务器发送类型为 32 的消息对象（参见代码7-59），用于获取当前用户当前年度的出勤数据。服务器从数据库中读取相应记录后，返回类型为 32 的消息给客户端，客户端 Socket 监听器收到这个类型为 32 的消息后，把收到的数据作为 CheckAttendance（）构造函数的参数，启动查看员工出勤界面。

CheckAttendance 类（参见代码 7-61）用于显示查看考勤界面以及根据用户的选择更新界面上的数据。其中代码 7-60 实现了在 ClientSocketListener 类增加对员工出勤数据的处理方法。

【代码 7-59】 RequestAttendanceData 类

```java
package attendance;
import java.io.IOException;
import java.io.ObjectOutputStream;
import java.util.Calendar;
import ManageClientConnServer.ClientSocket;
import common.Message;
    public class RequestAttendanceData{
    private String id;
    private String date[];
    public RequestAttendanceData(String id){
        this.id = id;
        Calendar cal = Calendar.getInstance();
        Integer year = cal.get(Calendar.YEAR);
        String yearS = year.toString();
        date = new String[24];
        for(int i = 0, k = 1; i < 24; i += 2, k++){
            date[i] = new String(yearS + "-" + ("" + k) + "-" + "1");
            if(k == 2 || k == 4 || k == 6 || k == 9 || k == 11){
                date[i + 1] = new String(yearS + "-" + ("" + k) + "-" + "30");
            }
            else{
                date[i + 1] = new String(yearS + "-" + ("" + k) + "-" + "31");
            }
        }
    Message ms = new Message();
    ms.setMesstype(32);
    ms.setSender(id);
    ms.setGetter(id);
    ms.setDate(date);
    try{
        ObjectOutputStream oos = new ObjectOutputStream(new ClientSocket()
    .getClientConnServer(id).getS().getOutputStream());
        oos.writeObject(ms);
        Thread.sleep(100);
    }catch(IOException e1){
```

```
            e1.printStackTrace();
        }catch(InterruptedException e){
            e.printStackTrace();
        }
    }
}
```

【代码 7-60】 修改 ClientSocketListener 类，增加对员工出勤数据的处理

```
public void run(){
...
//后续与服务器端交互信息的代码加在这里;
else if(ms.getMesstype() == 32){//各月份签到天数数据
        String id = ms.getSender();
        int sum[] = ms.getDay();
        ArrayList List = ms.getList();
        new CheckAttendance(id, sum, List);
    }
else if(ms.getMesstype() == 100){ //按员工 ID 查得的出勤数据
        String id = ms.getSender();
        int date[] = ms.getDay();
        ManageAttendance.addArray(id, date);
    }
else if(ms.getMesstype() == 102){ //按年份查得的出勤数据
        String id = ms.getSender();
        int date[] = ms.getDay();
        ManageAttendance.addArray(id, date);
    }
...
}
```

【代码 7-61】 CheckAttendance 类

```
package attendance;
import java.awt.Choice;
import java.awt.event.*;
import java.io.IOException;
import java.io.ObjectOutputStream;
import java.util.ArrayList;
import java.util.Calendar;
import javax.swing.*;
;import ManageClientConnServer.ClientSocket;
import ManageClientConnServer.ManageAttendance;
import common.Message;
public class CheckAttendance extends JFrame implements ActionListener, Item-
```

```java
Listener{
    private String id;
    private JLabel label[];
    private Choice ID;
    private JTextField field[];
    private JLabel labelId;
    private JLabel labelY;
    private Choice year;
    private JButton buttonD;
    private Calendar calendar;
    private int thisY;
    private String selected;
    public CheckAttendance(String id, int n[], ArrayList List){
        this.id = id;
        int sum[] = n;
        this.setSize(200,400);
        this.setLayout(null);
        calendar = Calendar.getInstance();
        thisY = calendar.get(Calendar.YEAR);
        labelId = new JLabel("ID");
        labelId.setSize(30,20);
        labelId.setLocation(10,0);
        ID = new Choice();
        ID.setSize(40,20);
        ID.setLocation(40,0);
        ID.addItemListener(this);
        for(int i = 0; i < List.size(); i++){
            ID.add(List.get(i) + "");
        }
        ID.select(0);
        selected = ID.getItem(ID.getSelectedIndex());
        this.add(labelId);
        this.add(ID);
        labelY = new JLabel("年份");
        labelY.setSize(30,20);
        labelY.setLocation(0,20);
        year = new Choice();
        year.setSize(60,20);
        year.setLocation(40,20);
        year.addItemListener(this);
        for(int i = 0, k = 2012; i < 10; i++, k++) {
            year.add(k + "");
```

```java
            }
            year.select(0);
            this.add(labelY);
            this.add(year);
            label = new JLabel[12];
            for (int i = 0; i < 12; i++){
                label[i] = new JLabel("" + (i + 1) + "月");
                label[i].setSize(30, 20);
            }
            field = new JTextField[12];
            for (int i = 0; i < 12; i++){
                field[i] = new JTextField("" + sum[i]);
                field[i].setSize(40, 20);
                field[i].setEditable(false);
            }
            for (int i = 0, k = 1, j = 2; i < 12; i++, j++){
                label[i].setLocation(k * 10, j * 20);
            }
            for (int i = 0, k = 1, j = 2; i < 12; i++, j++) {
            field[i].setLocation(k * 40, j * 20);
            }
            for (int k = 0; k < 12; k++) {
                this.add(label[k]);
                this.add(field[k]);
            }
            buttonD = new JButton("清空记录");
            buttonD.setSize(100, 30);
            buttonD.setLocation(50, 300);
            buttonD.addActionListener(this);
            this.add(buttonD);
            this.setLocationRelativeTo(null);
            this.setVisible(true);
        }
    @Override
    public void actionPerformed(ActionEvent e) { //删除考勤记录
            int select = JOptionPane.showConfirmDialog(null, "是否删除?", "确认窗口", JOptionPane.YES_NO_OPTION);
            if (select == JOptionPane.YES_OPTION){
                int select2 = JOptionPane.showConfirmDialog(null, "重新确认,是否删除?", "确认窗口", JOptionPane.YES_NO_OPTION);
                if (select2 == JOptionPane.YES_OPTION) {
                    Message ms = new Message();
```

```java
                ms.setMesstype(101);
                ms.setSender(selected);
                ms.setGetter(selected);
                try{
                    ObjectOutputStream oos = new ObjectOutputStream(new Client-
Socket().getClientConnServer(id).getS().getOutputStream());
                    oos.writeObject(ms);
                    Thread.sleep(100);
                }catch(IOException e1) {
                    e1.printStackTrace();
                } catch (InterruptedException er) {
                    e1.printStackTrace();
                }
                for(int i = 0; i < field.length; i++) {
                    field[i].setText("0");
                }
            }
        }
    }
    @Override
    public void itemStateChanged(ItemEvent e) {
    Message ms = new Message();
        if(e.getSource() == ID){
        selected = e.getItem().toString();
        ms.setMesstype(102); //选择其他ID时
        }
        else{
        int y = Integer.parseInt(e.getItem() + "");
        if (y > thisY) {
            for(int i = 0; i < field.length; i++) {
                field[i].setText("0");
            }
        }
        else {
            ms.setMesstype(100); //选择其他年份时
        }
    }
    String date[] = new String[24];
    String yearS = (thisY + "");
    for(int i = 0, k = 1; i < 24; i += 2, k++) {
        date[i] = new String(yearS + "-" + ("" + k) + "-" + "1");
        if(k == 2 || k == 4 || k == 6 || k == 9 || k == 11) {
```

```
                date[i + 1] = new String(yearS + "-" + ("" + k) + "-" + "30");
            }
            else{
                date[i + 1] = new String(yearS + "-" + ("" + k) + "-" + "31");
            }
        }
        ms.setSender(selected);
        ms.setGetter(selected);
        ms.setDate(date);
        try{
                ObjectOutputStream oos = new ObjectOutputStream(new ClientSocket()
            .getClientConnServer(id).getS().getOutputStream());
                oos.writeObject(ms);
                Thread.sleep(100);
        }catch(IOException e1){
            e1.printStackTrace();
            }catch(InterruptedException er){
            er.printStackTrace();
        }
        int day[] = ManageAttendance.getArray(selected);
        for(int i = 0; i < field.length; i++){
            field[i].setText(day[i] + "");
        }
        validate();
    }
}
```

（3）服务器端：实现考勤数据查看和删除功能。当收到客户端查看员工出勤记录时，根据指定的用户和指定的年份（参见代码 7-62），客户服务线程 ClientHandler 从数据库中读取所需数据，并返回给客户端（参见代码 7-63）；当收到清空出勤记录时，客户服务线程 ClientHandler 则删除指定用户和指定年份的出勤数据（参见代码 7-64）。

【代码 7-62】修改 ClientHandler 类，增加对类型为 32/100/101/102 消息的处理

```
public void run(){
    …
        else if(ms.getMesstype() == 32 || ms.getMesstype() == 102){  //各月份考勤情况
            String id = ms.getSender();
            String date[] = ms.getDate();
            int sum[] = new JDBC().attendState(id, date);
            System.out.println(sum.length);
            ArrayList List = new JDBC().IDSearch();
```

```java
            Message mA = new Message();
            mA.setMesstype(ms.getMesstype());
            mA.setSender(id);
            mA.setDay(sum);
            mA.setList(List);
            ObjectOutputStream oos = new ObjectOutputStream(s.getOutputStream());
            oos.writeObject(mA);
        }
        else if (ms.getMesstype() == 100) { //选择其他月份
            String id = ms.getSender();
            String date[] = ms.getDate();
            int sum[] = new JDBC().attendState(id, date);
            Message mA = new Message();
            mA.setMesstype(100);
            mA.setSender(id);
            mA.setDay(sum);
            ObjectOutputStream oos = new ObjectOutputStream(s.getOutputStream());
            oos.writeObject(mA);
        }
        else if (ms.getMesstype() == 101) { //删除某ID全部记录
            String id = ms.getSender();
            String result = new JDBC().deleteAllAttend(id);
            Message m = new Message();
            m.setMesstype(101);
            m.setContent(result);
            ObjectOutputStream oos = new ObjectOutputStream(s.getOutputStream());
            oos.writeObject(m);
        }
        ...
    }
```

【代码7-63】修改JDBC类,增加按ID和年份查看出勤情况的attendState()方法

```java
public int[] attendState(String id1, String string[]){
    String id = id1;
    String date[] = string;
    int number = 0;
    int sum[] = new int[12];
    String string1 = new String();
```

```java
try{
    Class.forName(driver);
    conn = DriverManager.getConnection(url, userName, sqlpassword);
    System.out.println("数据库连接成功");
    for (int i = 0, k = 0; i < 12; i++, k += 2) {
        stmt = conn.createStatement();
        number = 0;
        String sql = "select * from attend where time between'" + date[k]
            + "' and '" + date[k + 1] + "'";
        rs = stmt.executeQuery(sql);
        while (rs.next()){
            string1 = rs.getString(1);
            if (string1.equals(id)){
                number++;
            }
        }
        sum[i] = number;
    }
    rs.close();
    stmt.close();
    conn.close();
}catch (ClassNotFoundException e) {
    e.printStackTrace();
} catch (SQLException e) {
    e.printStackTrace();
}
    return sum;//返回今天是否签到了
}
```

【代码 7 - 64】修改 JDBC 类，增加查询 ID 和删除出勤记录的 IDSearch（ ） 和 deleteAllAttend（ ） 方法

```java
public ArrayList IDSearch(){ //得到各个 ID
    ArrayList<String> List = new ArrayList<String>();
    try{
        Class.forName(driver);
        conn = DriverManager.getConnection(url, userName, sqlpassword);
        System.out.println("数据库连接成功");
        stmt = conn.createStatement();
        String sql = "select id from userInfo";
        rs = stmt.executeQuery(sql);
        while (rs.next()){
            List.add(rs.getString(1));
        }
    }
```

```java
        rs.close();
        stmt.close();
        conn.close();
    }catch(ClassNotFoundException e){
        e.printStackTrace();
    }catch(SQLException e){
        e.printStackTrace();
    }
    return List;  //返回各个ID
}

public String deleteAllAttend(String id){  //删除考勤记录
    try{
        Class.forName(driver);
        conn = DriverManager.getConnection(url, userName, sqlpassword);
        stmt = conn.createStatement();
        String sql = "delete from attend where id = '" + id + "'";
        stmt.executeUpdate(sql);
    }catch(ClassNotFoundException e){
        e.printStackTrace();
    }catch(SQLException e){
        e.printStackTrace();
    }
    return "Yes";
}
```

7.4 总结与回顾

通过以上章节的增量项目开发，最终实现了办公管理系统的预定项目目标。本章实现的 C/S 结构的办公管理系统可方便地在局域网内使用，具有很好的性能和一定的应用价值。然而，本系统在代码的实现上仍然存在以下问题，由于篇幅有限，本章不作详细分析和实现，读者可根据自身情况对以上实现的办公管理系统进行修订，以期达到更好的实现效果和应用价值：

（1）客户端在向服务器端发送信息请求后需要 sleep 一段时间，才能收到数据。然而，实际应用中，我们并不是很清楚服务器需要多长事件才能返回信息。该问题在用户量比较大、用户访问比较频繁的情况下更为突出。解决方法：可以使用线程间的互斥技术解决该问题，具体实现方法可参考《Java 程序设计及应用——增量式项目驱动一体化教程》中局域网聊天器客户端的实现过程。

（2）客户端保存了用户自己的权限，且在客户端编写代码来判断用户的访问权限以决定该用户是否可以执行一些特定操作。这样的权限管理存在安全漏洞，如果存在一些恶意程序，则可以利用这些漏洞执行一些非法操作。解决方法：将用户是否具有执行某个操作的判定放在服务器端执行,这样不管客户端有没有被恶意使用,都尽可能保证了系统的安全性。

(3) 在程序设计的过程中,没有考虑到对异常的处理,导致出现异常后,系统没有办法恢复,系统的可靠性较差。解决方法:在出现异常后,考虑操作的恢复或系统状态恢复的实现。

(4) 相对于服务器来说,客户端执行了大量的业务操作,导致客户端代码比较多,使得本末倒置。解决方法:从系统架构上重新设计服务器和客户端的功能分布。

可以通过解决以上问题提高办公管理系统的软件质量和可靠性。所以,在软件项目的开发过程中,一定要做好项目需求、考虑软件的可靠性、设计比较合理的软件架构,最终达到好的用户体验。

7.5 软件使用说明

7.5.1 开发环境

(1) Java 集成开发环境:MyEclipse 8.5。

(2) 数据库管理系统:MySQL。

(3) 数据库配置步骤:①在 MySQL 中创建 office 数据库;②依次创建 userinfo、publication、message、groups 和 attend 等 5 张数据表;③用户表 uerinfo 中创建经理用户(经理用户权限为"mgr",一般用户权限为"user"),也可以直接使用下面代码 7-46 中的 T-SQL 脚本完成数据库配置。

【代码 7-65】office 数据库脚本

```
create database office;
use office;
create table userinfo(
    id VARCHAR(16) primary key,
    authority VARCHAR(10) NOT NULL,
    pass VARCHAR(16) NOT NULL,
    income int NOT NULL,
    age int NOT NULL,
    sex VARCHAR(4) NOT NULL,
    address VARCHAR(100) NOT NULL,
    telpNum VARCHAR(15) NOT NULL,
    QQNum VARCHAR(15) NOT NULL
);
insert into userinfo values('admin','mgr','admin',5000,33,'m','广州','61787330','123456');
insert into userinfo values('niuniu','mgr','user',3500,20,'m','广州','61787331','234567');
insert into userinfo values('陈牛牛','0','******',4000,30,'m','广州',
```

```sql
'161787331','654321');
    insert into userinfo values('优优','0','******',4000,30,'m','广州','161787331','654321');
    create table message(
        SenderID VARCHAR(16) primary key NOT NULL,
        GetterID VARCHAR(16) NOT NULL,
        Cont VARCHAR(150) NOT NULL
    );
    create table groups(
        id VARCHAR(16) primary key not null,
        groupName VARCHAR(40) not null
    );

    create table publication(
    title VARCHAR(40) primary key,
    message TEXT not null,
    time timestamp default current_timestamp on update current_timestamp
    );
    create table attend(
    id INT primary key not null,
    userid VARCHAR(16) not null,
    time timestamp default current_timestamp on update current_timestamp,
    present VARCHAR(10) null
    );
```

7.5.2 软件使用方法

（1）软件运行环境：要求服务器和客户端已经安装 JDK。

（2）软件部署：①服务器端：运行（由 MyEclipse 导出的）可执行文件 officeServer.jar；②客户端：在与客户端程序 officeClient.jar 同目录下，创建一个文件 server.cfg，文件中填写服务器端的 IP 地址，然后运行 officeClient.jar 即可。

7.6 本章小结

本章设计并实现了一个办公管理系统，在章节安排上首先对办公管理系统进行需求分析，确定系统的目标；然后进行功能分析和软件设计，确定系统的增量开发计划；最后按照增量开发的方法，按照设计的增量分阶段实现每个增量要求的目标。

办公管理系统共实现了四个增量：搭建系统主体架构、实现员工管理功能、实现公告和实现考勤功能，每个增量使用的技术和案例模版可参考专项实训部分。在功能分析和软件设计部分介绍的通信功能，本章没有详细介绍，有兴趣的读者可参考本教材案例资源包。

第8章　Web考勤系统

与以往单纯的Java项目不同，本章的实训项目将采用Web编程技术JavaScript，结合服务器Tomcat和MySQL数据库，实现一个基于Web的考勤系统。通过本项目的实现，期望给读者以Web项目的形式展现Java Web编程相关的技术，使读者对Java技术的应用有更深入的了解。

该Web考勤系统主要面向学校教务管理人员、教师等用户，是一个为学校各用户提供服务的综合管理系统。本系统重点对考勤系统中的在线记录考勤状态、在线打印考勤状态、自动获得上课时间表、自动获得学生名单、多方面应用共享登录信息等功能进行设计与实现。本系统可以取代传统的手工考勤和统计方式，教务管理人员、教师等用户可以很方便地完成学生考勤和考勤统计工作。

8.1　需求分析与项目目标

8.1.1　用例分析

Web考勤系统需要实现的功能包括：在线记录考勤状态、在线打印考勤状态、自动获得上课时间表、自动获得学生名单、多方面应用共享登录信息等，其主要用户有管理员、教师和学生。管理员拥有系统中的所有权限，教师则可以对所负责的课程进行考勤、考勤修改、课程考勤情况统计等权限，而学生则可以查看自己的考勤记录。

经过分析，我们可以采用图8-1所示的用例图来概括本系统的使用场景。首先，所有用户必须在登录后，才能获取相应用户权限，进而访问考勤系统的相应功能。不同用户在考勤系统中所拥有的权限如下所示：

（1）所有用户共享权限：登录、查看个人基本信息、修改密码。
（2）学生：查看个人考勤情况。
（3）教师：增加授课、修改授课、删除授课、获取班级信息、获取学生信息、查看考勤信息、修改考勤信息、查看学生考勤记录、输入规则获取成绩。
（4）管理员：添加用户、批量添加用户、删除用户、查看用户信息、修改用户信息；添加单元、修改单元、删除单元；添加课程、修改课程、删除课程。

下面，我们用表8-1来详细概括图8-1中的用例，并对每个用例进行具体描述。

图 8-1 Web 考勤系统用例

表 8–1 Web 考勤系统用例表

用例编号	用例名称	用例描述
用例 1	登录	共同权限：登录、判断用户是否存在、获得用户头像、验证用户密码是否正确
用例 2	查看用户信息	共同权限：查看用户详细信息
用例 3	修改密码	共同权限：修改当前用户密码
用例 4	获得课程表	共同权限：通过课程的周数、星期数、上课时间、下课时间、相关班级、授课地点等，按照时间安排统计并显示当前用户所有绑定的课程，同时能智能显示即将到来的课程表
用例 5	查看班级信息	教师权限：查看某一个班级的学生信息
用例 6	修改考勤信息	教师权限：修改考勤信息
用例 7	查看考勤信息	教师权限：查看考勤信息
用例 8	新建授课	教师权限：选择相应的时间、地点和班级
用例 9	修改授课	教师权限：修改相应的时间、地点和班级
用例 10	删除授课	教师权限：判断是否存在数据关联
用例 11	获得学生名单	教师/管理员权限：获得当前课程绑定的学生班级的所有名单，并且根据学生数目生成快捷链接
用例 12	新建用户	管理员权限：新建用户、判断用户名是否重复、判断用户信息是否正确、判断班级是否重复、判断用户角色是否重复
用例 13	修改用户	管理员权限：判断用户名是否重复、判断用户信息是否正确、判断班级是否重复、判断用户角色是否重复
用例 14	删除用户	管理员权限：判断用户是否有其他记录
用例 15	新建课程	管理员权限
用例 16	修改课程	管理员权限
用例 17	删除课程	管理员权限
用例 18	批量增加用户	管理员权限
用例 19	新建节	管理员权限：新建一节课
用例 20	修改节	管理员权限：修改一节课的信息

8.1.2 业务流分析

在用户登录前，用户为未认证用户，未认证用户必须登录后才能使用 Web 考勤系统。未认证的用户访问系统时，系统将会默认显示登录页，用户必须通过自己的账号和密码来登录。登录页面采用 Ajax 技术进行登录验证，将对用户的用户名，用户名加上密码进行验证，验证后在登录页面通过提示进行提醒。

如果用户登录超时，则提示用户登录超时；如果用户名或者密码错误，则提示用户账号或密码错误；如果登录成功，考勤系统则将对该用户进行记录，并根据用户的类型（学生、教师、管理员），分别显示学生界面、教师界面、管理员界面。图8-2是Web考勤系统总体业务流，图8-3至图8-6分别是未认证用户、管理员、教师和学生的数据业务流。

图8-2　Web考勤系统业务流

图8-3　未认证用户业务流程

图8-4　管理员业务流程

管理员可以对用户、单元、课程进行操作。以管理员身份登录后，用户将会进入管理员首页，管理员可以获取所有的基础数据，包括班级、学生、状态、节、课程名称等。管理员拥有对所有基础数据的修改权限。

以教师身份登录后，系统将进入教师首页，教师可以获得相应的班级信息，获得其授课班级的所有学生的信息，以及获得各种基础信息，如课程、课程表、班级。教师用户拥有获取上课的学生名单，创建、修改上课时间安排，进行考勤，读取任课班级的相应课程考勤信息等权限。

图 8-5 教师业务流程

图 8-6 学生业务流程

以学生身份进行登录后，将进入学生首页，这里只会获取学生用户的数据，如个人课程表、个人考勤记录、个人请假等记录。学生用户只能查询个人的数据，无法对其他数据进行查看，如其他班级、其他学生的数据。

8.2 界面设计

8.2.1 用户登录界面

实现用户登录界面采用的技术是单点登录（Single Sign On，简称为 SSO）。SSO 是目前比较流行的企业业务整合的解决方案之一。SSO 的定义是在多个应用系统中，用户只需要登录一次就可以访问所有相互信任的应用系统。

1. 技术实现机制

当用户第一次登录系统中任何一个子系统时，系统将会引导用户到认证系统中，根据用户提供的信息进行认证。通过认证后，将用户引导至认证前的页面并可以继续访问需要相应权限的功能，同时服务器将该用户的认证信息存储到相应的位置，当用户访问其他子系统时，如果通过浏览器发送的信息和服务器的认证信息相匹配，即可实现统一登录。

2. 技术实现优势

用户只需要一次登录即可访问系统中所有的子系统，这样不仅减少了用户在使用系统的过程中浪费的时间，而且对于开发人员来说，单点登录可以降低系统开发的难度，可以将系统切分成多个子系统分别进行开发，最后再进行系统整合就可以了，而且也方便系统的横向拓展，方便开发人员进行功能的添加和删除。

我们需要实现如图8-7所示的考勤系统首页——系统统一登录界面。

图8-7 用户登录界面

系统要求用户在图8-8所示的账号输入框内输入账号，登录页面账号输入框将会自动判断是否输入正确的用户名。当用户什么都不输入或者输入不合法字符后跳向密码输入框时将会提醒用户没有输入或输入错误。这时只是本地进行验证而已，没有向服务端进行验证。这时，我们只需要重新激活输入框便可以取消错误提示，重新输入正确的数据即可，如果再次输入错误信息将继续进行提示。

当页面存在错误提示框时，【登录】按钮将不会执行登录操作，必须填写上正确的信息方可进行登录操作。

图8-8 登录页面的账号输入框-初始界面

当用户输入可以使用的字符后跳到密码框时，页面将会发送账号到服务器那边进行验证，验证的过程中并不影响用户的其他操作，如输入密码等。

验证完毕后会在页面进行提示，如果该账号并不存在，将会弹出提示账号不存在（如图8-9所示），如果账号已经存在了，则会动态修改个人头像。

当账号输入完成后，我们将输入密码，然后进行登录，在单击【登录】按钮后，系统将会向服务器验证账号和密码的正确性。

图 8 – 9　账号输入框 – 账号不存在

因为在账号输入框中已经确定了账号的正确性，所以在这里只需要对该账号的密码进行验证，验证结果将会在页面进行显示。

如果密码错误，则会在页面上提示密码出错（如图 8 – 10 所示），必须重新激活密码输入框并重新输入密码，如果再次输入错误信息将继续进行提示。如果输入正确，则后台会判断相应的权限并跳转到相应的页面。

图 8 – 10　密码框 – 密码错误

8.2.2　学生界面

学生用户界面包括四个功能模块：

1. 查看个人信息

通过查看个人信息功能，学生可以在网上校对个人信息并随时修改自己的密码，当信息出现问题时，可以向学校提出修改信息的请求。

2. 查看课程表

通过课程设置，系统可以自动将课程转换成课程表，学生可以查看自己的课程表。系统主要以班级为单位提供课程表。

3. 查看考勤记录

登录后，学生可以查看自己的考勤记录，可以通过选择课程和第几周来准确获得考勤记录，也可以获得自己的所有考勤记录列表。

8.2.3　教师界面

教师用户界面包括四个功能模块：

1. 查看个人信息

通过查看个人信息功能，教师可以在网上校对个人信息并修改自己的密码，当教师信息出现错误时，可以向学校提出修改信息的请求。

2. 课程表管理模块

教师必须设置个人的课程表，将课程和上课时间还有上课地点以及相应的班级进行设置即可，设置完成后将会自动生成个人的课程表，教师对自己的课程表可以进行任意修改。

3. 考勤模块

教师在上课时间可以进行课堂考勤，考勤时只需对学生的考勤状态进行标记即可。教师还可以查看考勤记录，包括整个班级的考勤记录、每个学生的考勤记录、请假列表等。

4. 下载模块

教师端可以下载 Excel 格式的考勤记录表，还可以通过设置单项分数自动计算课堂考勤总分数。

8.2.4 管理员界面

管理员用户界面包括六个功能：

1. 查看个人信息

通过查看个人信息，管理员可以在网上校对个人信息并修改自己的密码，当信息出现错误时，可以进行修改。

2. 角色管理

在角色管理模块中，可以区分各种用户角色，本系统提供三种角色，分别为管理员、教师和学生，不同的角色将有不同的用户界面和功能。

3. 用户管理

管理员对用户拥有完整的管理权限，包括增加、修改和删除等操作。

4. 班级管理

管理员对班级拥有完整的管理权限，包括增加、修改和删除等操作。

5. 课程名称管理

管理员对课程名称拥有完整的管理权限，包括增加、修改和删除等操作。

6. 查看操作记录

查看管理员自己在系统上的操作记录。

8.2.5 界面效果

1. 个人信息界面

学生、教师和管理员都可以查看个人信息，但只有管理员具有修改权限（如图 8-11 所示）。

图 8-11　个人信息界面

2. 管理员界面

管理员界面（如图 8-12 所示）包括用户管理、单元管理、课程管理、权限管理、状态管理，每个子模块下面都包括相应的增、删、查、改功能。

图 8-12　管理员界面

在管理员界面左侧，是管理员界面菜单栏（如图 8-13 所示），通过菜单栏，管理员可以进行用户管理、单元管理、课程管理、权限管理、状态管理等操作。

（1）用户管理。用户管理是对相应用户的添加、修改、删除操作。单击用户管理将会弹出用户选择筛选面板（如图 8-14 所示），该面板提供了根据角色和根据班级进行选择两种筛选器。

通过"角色选择"将会对符合类型的列表项进行显示，不符合的项进行隐藏，这

图8-13 管理员菜单栏

图8-14 用户选择筛选面板

样可以方便对用户进行管理。

通过"班级选择"将会发送数据请求服务器获取相关用户数据,由于班级是以无限级数的数据结构形式设计的,所以理论上我们的界面将会显示一个无限级的菜单,但是根据实际的条件我们知道,任何单位都不会有一个无限级的部门列表,所以在显示这方面我们使用一个"多级抽屉"来显示班级之间的关系,顶级班级将是最大的抽屉,单击下去将是一棵班级树。

我们假设系统运行在一个由系、专业、班级组成的组织部门中,这样,菜单将会有三级,即系别、专业、班级,人员均会存放在最底层的单位中,所以系下面没有相应的人员,专业下面也没有相应的人员显示,只有班级下面会绑定相应的人员。

我们在单击电子系后将会看到电子系的抽屉打开,下面会有电子系包含的所有下级菜单,也就是下级的所有专业【如图8-15(a)所示】,同样在专业下面会包含相应的班级【如图8-15（b）所示】。

（a）　　　　　　　　　　　　（b）
图8-15　用户选择筛选面板——根据班级筛选

由于菜单始终置顶并且固定于相对位置的，所以菜单采用的策略是只显示"单一抽屉"，即不能同时打开两个专业的班级列表，当打开另外一个专业时，之前打开班级列表抽屉将会自动闭合，这样既节省空间，又能够最明显地显示相关信息（如图8-16所示）。由于菜单使用了置顶并且固定位置，所以我们必须有功能按钮能够自由打开/关闭菜单，这样就能最大限度地使用空间和方便用户操作。

当单击按钮后会隐藏菜单，这时按钮指针指向外侧，提示操作人员这里可以向外面展示更多功能。同样，打开菜单后，按钮的指针将会指向内侧，提示操作人员可以对菜单进行闭合操作（如图8-16所示）。

图8-16　单一抽屉式筛选面板

单击任意一个班级后，系统将显示如图8-17所示的班级管理筛选界面，隐藏筛选栏即可看到整个人员列表（如图 8-18 所示）。

图 8-17 班级管理界面

图 8-18 人员列表

1）添加用户。单击图 8-18 中的管理员菜单栏中的"添加用户"选项，系统会弹出"添加用户"对话框（如图 8-19 所示）。此对话框提供的用户信息项分别为头像、学号、姓名、性别、班级、电话、短号、用户角色等，其中学号不允许重复，姓名必须是 2~4 个中文字符，性别必须选择，默认为男，班级必须选择，而且不能重复选择，班级可以多个，但是每个班级只能选一次，电话为 11 位手机号码，短号为 5~6 位手机号码，用户角色必须选择，而且不能重复选择，角色可以多个，但是每个只能选择一次。

当操作焦点（闪烁光标）离开学号框时，会自动进行验证，判断学号的正确性。如果用户学号格式正确，系统会与后台验证来判断学号是否存在。当学号格式不正确或者已经存在时，输入框会显示红边以提醒管理员，此红边在用户选择学号框时会自动消失，表示可以修改学号，学号修改完成后系统会重新进行判断（如图 8-20 所示）。

图 8-19 添加用户对话框

图 8-20 用户学号出错

当输入正确的学号后，系统会为学号框添加绿色边框，表示学号输入正确。

性别选项是一个开关按钮，管理员通过单击或者拖动该按钮，可以选择另外一个选项（如图 8-21 所示）。其中蓝色的选项表示已选中，白色表示没有选中。

211

图 8-21　性别选项

在选择班级时，系统在可选列表中显示所有符合条件的班级，管理员只需要选择后单击【添加】按钮即可添加所属班级（如图 8-22 所示），同一个班级不能重复添加。每添加一个班级，系统都会在前台判断班级是否重复选取，如果重复选取，系统会出现警告框提示和窗口提示，提示管理员选择其他班级或者退选。

图 8-22　班级选项

角色添加选项（如图 8-23 所示）的实现与添加班级选项类似，管理员必须按照规则选择角色，然后单击【添加】按钮即可添加到角色列表里面。当添加的角色已经选取时，系统便会提示，提示方法有弹出提示框和警告框（如图 8-24 所示），提示管理员选择其他班级或者退选。

图 8-23　角色选项

图 8-24　角色重复选择提示

角色退选：将鼠标移动到已选角色列表中的角色上，选中的角色项将变为灰白色（如图 8-25 所示），单击即可取消该项。

图 8-25 角色退选

上述信息填写或者选择完成后,如果所有项都填写完整并且检验合格(如图 8-26 所示),管理员即可通过单击【保存】按钮来完成用户添加。用户添加成功后,系统将会进行提示(如图 8-27 所示),这时我们可以在相应班级中找到该用户(如图 8-28 所示)。

图 8-26 完成用户添加

图 8-27 用户添加成功

图 8-28 新添加的用户

2）修改用户信息。管理员选中某个用户后单击"修改"选项，系统将会在添加框中先预填该用户的基础信息（如图8-29所示），此时就可以根据这些基础信息进行修改。预填的信息为用户的基本信息，对其修改后单击【保存】按钮，或者【关闭】按钮。

图8-29 用户信息修改

（2）班级管理。班级管理是对相应班级的添加、修改、删除操作。单击"班级管理"后，系统将显示如图8-30所示的班级管理界面。选择班级后将向服务器请求访问相应的人员数据，人员数据包括序号、学号、姓名、性别、电话、短号和修改等项。

图8-30 班级数据

在班级管理中，班级出现在一棵树的叶子节点上（如图8-31所示），以缩进的方式显示出根和子节点，可以对每个节点单元进行添加和修改。

图 8-31　班级树

（3）课程管理。课程管理是对相应课程的添加、修改、删除操作。课程列表（如图 8-32 所示）中列出了课程名称、是否启用等项，可对课程进行修改。

图 8-32　课程管理

添加课程界面（如图 8-33 所示），可以添加课程的名称和是否启用。

图 8-33　添加课程

（4）节管理。节管理是对相应节的添加、修改、删除操作。通过节管理，管理员可以创建一节课，修改一节课的信息。每节课的信息有名称、开始时间、结束时间等项（如图 8-34 所示）。

图 8-34　节管理

通过"添加节"界面（如图 8-35 所示），可以填写名称、选择开始时间、选择结束时间等操作。

图 8-35　添加节

（5）出勤状态管理。出勤状态管理是对相应状态的添加、修改、删除操作。出勤状态管理中默认包括"早退"、"缺勤"、"已到"、"请假"和"迟到"五个状态，每个状态有名称、是否启用等信息，管理员可以通过"修改"选项进行修改（如图 8-36 所示）。

添加状态界面（如图 8-37 所示），可以填写状态名称和是否启用等操作。

图 8-36 出勤状态管理

图 8-37 出勤状态添加

(6) 角色管理。角色管理是对相应角色的添加、修改、删除操作。通过角色管理(如图 8-38 所示),系统提供了角色名称、权限和是否启用的设置方法。

图 8-38 角色管理

（7）系统设置界面。通过系统设置界面，可以设置考勤系统的相关信息（如图 8 - 39 所示）。

图 8 - 39　系统设置界面

（8）系统连接界面。通过系统连接界面，可快速访问其他角色的页面（如图 8 - 40 所示）。

图 8 - 40　系统连接界面

3. 教师导航条

教师可以通过导航条（如图 8 - 41 所示）使用本系统的各项功能，包括查看课程表、查看学生列表、考勤、查看考勤记录等。

图 8 - 41　教师导航条

4. 课程表界面

课程表界面（如图 8 - 42 所示）分为教师和学生两种，将显示不同的课程表数据。

5. 考勤界面

本系统提供两种不同的考勤界面供教师选择使用。考勤界面 1（如图 8 - 43 所示）以比较大的字体显示单个学生的信息，界面下面的五个按钮对应不同的出勤状态。考勤界面 2（如图 8 - 44 所示）同时显示多个学生列表，并在每个学生名字下面显示五个按钮，对应不同的出勤状态。

6. 查看考勤记录界面

教师可以通过选择课程和班级来查看考勤记录，教师可以查询班级的记录（如图

图 8-42　课程表界面

图 8-43　考勤界面 1

8-45 所示），也可以查看单个学生的出勤记录（如图 8-46 所示），还可以对班级考勤记录进行统计和 Excel 表格下载。学生只能查看自己的考勤记录。

8.2.6　界面设计技术——响应式设计

当前，几乎每个用户都希望网站服务器能提供不同移动版本的网页。最完美的情况，就是为 iPhone、iPad、黑莓、Kindle 等客户端各自打造一款，页面分辨率必须兼容任何设备的网页。这样，我们将无尽地开发出各种兼容版本，因为随着移动设备的更新换代，屏幕以及分辨率的大小都会改变，而我们的网页又不能适应各种变化。对于多数

图 8-44 考勤界面 2

图 8-45 班级考勤记录界面

图 8-46 个人考勤记录界面

网站来说，为每种新设备及分辨率创建其独立的版本根本就是不切实际的；结果就是，我们将会赢得使用某些设备的用户群，而失去那些使用其他设备的用户群。

响应式 Web 设计（Responsive Web design）的理念是，页面的设计与开发应当根据用户行为以及设备环境（系统平台、屏幕尺寸、屏幕定向等）进行相应的响应和调整。具体的实践方式由多方面组成，包括弹性的网格、布局、图片和 CSS media query 的使用等。无论用户正在使用笔记本还是 iPad，我们的页面都应该能够自动切换分辨率、图片尺寸及相关脚本功能等，以适应不同设备；换句话说，页面应该有能力去自动响应用户的设备环境。这样，我们就可以不必为不断到来的新设备做专门的版本设计和开发。

1. 响应式设计方法

采用响应式设计的网站使用 CSS3 Media queries，即一种对 @ media 规则的扩展，以及流式的基于比例的网格和自适应大小的图像以适应不同大小的设备：

Media queries 允许网页根据访问站点设备的特点而使用不同的 CSS 样式规则，最常用的是浏览器的宽度。

流式网格概念要求页面元素使用相对单位如百分比或 EM 调整大小，而不是绝对的单位如像素或点。

CSS3 支持 CSS2.1 所支持的所有媒体类型，如 screen、print、handheld 等，同时添加了很多涉及媒体类型的功能属性，包括 max–width（最大宽度）、device–width（设备宽度）、orientation（屏幕定向、横屏或竖屏）和 color。在 CSS3 发布之后出现的新产品，如 iPad 或 Android 相关设备，都可以完美地支持这些属性。所以我们可以通过 media query 为新设备设置独特的样式，而忽略那些不支持 CSS3 的台式机中的旧浏览器。

CSS3 专有的 media queries 功能来创建响应式 Web 设计（如代码 8–1 所示），可以更快更方便地实现，min–width 和 max–width 这两个属性很靠谱。通过 min–width 的设置，我们可以在浏览器窗口或设备屏幕宽度高于这个值的情况下，为页面指定一个特定的样式表；max–width 则反之。

【代码 8–1】

```
/* Smartphones (portrait and landscape) ----------- */
@ media only screen
and (min-device-width : 320px)
and (max-device-width : 480px) {
/* Styles */
}

/* Smartphones (landscape) ----------- */
@ media only screen
and (min-width : 321px) {
/* Styles */
}
```

```
/* Smartphones (portrait) -----------*/
@ media only screen
and (max-width: 320px) {
/* Styles */
}
```

代码 8-2 对各种设备进行判断,并得到最佳的样式。从大屏幕(分辨率大于 1200px)、普通屏幕(分辨率小于 1200px 大于 980px)、横放的平板电脑(分辨率小于 980px 大于 768px)、普通手机和平板电脑(分辨率小于 768 大于 480px)、普通手机(分辨率小于 480px),页面上的每个元素可以在每个判断中进行自定义,在页面载入的时候会自动更换分辨率的大小选择相应的样式。

【代码 8-2】

```
@ media (min-width: 1200px) {

}
@ media (min-width: 980px) and (max-width: 1199px){

}
@ media (min-width: 768px) and (max-width: 979px){

}
@ media (min-width: 481px) and (max-width: 767px){

}
@ media (max-width: 480px){

}
```

8.3 软件设计

8.3.1 系统架构设计

我们采用四层结构来构建 Web 考勤系统(如图 8-47 所示),由下到上分别为用户接口(Interface)层、应用(Application)层、领域(Domain)层和基础设施(Infrastructure)层。

用户接口层包含与其他系统进行交互的接口和通信设施,是与其他系统发生信息交互的接口,其主要技术方法有 Servlet、Java 远程方法调用(RMI)等。该层主要由 Facade、DTO、Assembler 三类组件构成。

应用层只是很薄的一层服务,其中主要的组件是 Service,所有的 Service 只负责协调并委派业务逻辑给领域对象进行处理,其本身并未真正实现业务逻辑,绝大部分的业务逻辑都由领域对象承载和实现。应用层会和许多其他组件进行交互,包括其他 Service、领域对象、Repository（仓储）。

领域层是整个系统的核心层，该层维护一个使用面向对象技术实现的领域模型，几乎全部的业务逻辑会在该层实现。Domain 层包含 Entity（实体）、Value Object（值对象）和 Repository（仓储）等多种重要的领域组件。

基础设施层为 Interfaces、Application 和 Domain 三层提供支撑。所有与具体平台、框架相关的实现会在 Infrastructure 中提供，避免其他三层特别是 Domain 层掺杂进这些实现过程中，从而"污染"领域模型。Infrastructure 中最常见的一类设施是对象持久化。

图 8 -47　Web 考勤系统架构

8.3.2　基础设施层

1. 主要的类图

（1）数据库辅助类（如图 8 - 48 所示）主要的作用是获取数据库链接并保持有一定量的数据库连接池，从而达到快速连接数据库的需要，其中连接池使用了 C3P0 这个类库，可以在配置文件中对相应的参数进行设置，也可以选择是否使用连接池。通过 getConnection（）方法获得 Connection 实例。

（2）配置文件辅助类（如图 8 - 49 所示）主要提供了读取配置文件的功能，主要是数据库连接的配置信息,包括C3P0连接池的配置信息。

图8-48 数据库链接辅助类

图8-49 配置文件辅助类

(3)序列化辅助类(如图8-50所示)可以将内存中的对象进行序列化到硬盘和从硬盘反序列化到内存中,可以保持对象的状态,通过接口提供的方法使用。

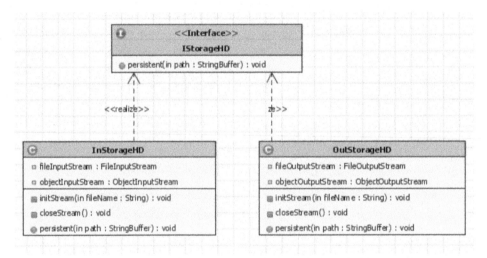

图8-50 序列化辅助类

(4)持久化辅助类(如图8-51所示)是通过对Model的操作将其持久化到数据库中。通过接口中的doModel()方法来实现,子类必须实现超类中声明的initExportSql()方法来获得SQL语句,并且实现doModel()方法将数据保存到数据库。

(5)资源库管理类(如图8-52所示)通过持有一个保持着资源库的Map来对资源库进行管理,注意该Map容器使用到了JDK1.5中新的API,一个线程安全的Map。所以在多线程的环境下可以保证对资源库的操作是同步的,可以通过方法putIfAbsent、remove、replace的原子性来保证。

ConcurrentMap常用的方法:

1)putIfAbsent(K key, V value):如果key已经不再与某个值相关联,则将它与value关联,该操作是原子的。

图 8-51 持久化辅助类

图 8-52 资源库管理类

2) remove (Object key, Object value)：只有 key 的条目映射到给定值时，才移除该键的条目，该操作是原子的。

3) replace (K key, V value)：只有目前 key 的条目映射到某一值时，才替换该键的条目为 value，该操作是原子的。

4) replace (K key, V oldValue, V newValue)：只有目前 key 的条目映射到 oldValue 时，才替换该键的条目为 newValue，该操作是原子的。

5) 通过对SQL语句的封装类（如图8-53所示），可以获取每个操作的SQL语句，从而方便数据库存取。

图8-53　SQL语句封装类

（6）通过一个事件队列（如图8-54所示）持有用户所有请求的事件，队列使用了BlockingQueue，在获取元素时等待队列变为非空，以及存储元素时等待空间变得可用。BlockingQueue实现是线程安全的。所有排队方法都可以使用内部锁或其他形式的并发控制来自动达到它们的目的。

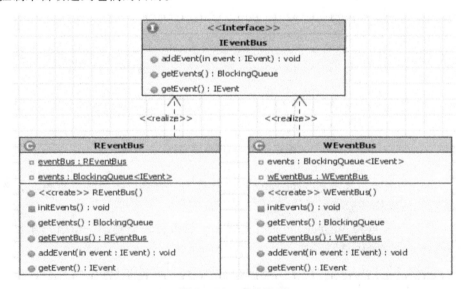

图8-54　事件队列

BlockingQueue 常用操作方法：

1) add（anObject）：把 anObject 加到 BlockingQueue 里，如果 BlockingQueue 可以容纳，则返回 true，否则抛出异常。

2) offer（anObject）：表示如果可能的话，将 anObject 加到 BlockingQueue 里，即如果 BlockingQueue 可以容纳，则返回 true，否则返回 false。

3) put（anObject）：把 anObject 加到 BlockingQueue 里，如果 BlockingQueue 没有空间，则调用此方法的线程被阻断直到 BlockingQueue 里有空间再继续。

4) poll（time）：取走 BlockingQueue 里排在首位的对象，若不能立即取出，则可以等 time 参数规定的时间，取不到时返回 null。

5) take（）：取走 BlockingQueue 里排在首位的对象，若 BlockingQueue 为空，阻断进入等待状态直到 BlockingQueue 有新的对象被加入为止。

通过事件（如图 8-55 所示）进行数据传递，事件接口提供了 doThings 和 notifyResource 方法，分别是事件的执行和回调接口。通过事件组件，可以降低系统的耦合，各松散耦合的组件和服务之间可以通过事件来进行数据传递。

2. 方法思想

（1）单例模式。Warehouse 仓库类（代码 8-3）和 EventBus 事件队列在系统中都只能有一个实例，所有的服务都会使用到同一个实例。单例模式有两种：懒汉模式、饿汉模式。懒汉模式是时间换空间，用到了才进行实例化，而饿汉模式是采用空间换取时间，一开始直接实例化。对于项目来说，上面两个类都是必须的，也就是说，两个类都会被实例化。本项目我们使用了饿汉模式，这样还可以避免懒汉模式的缺点，解决高并发时可能会出现多次实例化。

【代码 8-3】WareHouse 类

```
public class WareHouse{
private static WareHouse wareHouse = new WareHouse();
…；
}
```

（2）面向接口编程。接口是一组规则，规定了下面的实现必须基于该原则。在设计的过程中，由于处理业务的逻辑时会有不同的实现，所以统一接口（代码 8-4、代码 8-5），然后下面分不同的实现。

【代码 8-4】IEventBus 接口

```
public interface IEventBus{
    public void addEvent(IEvent event);
    public BlockingQueue<IEvent> getEvents();
    public IEvent getEvent() throws InterruptedException;
}
```

【代码 8-5】REventBus 类

```
public class REventBus implements IEventBus{
    private static REventBus eventBus = new REventBus();
    …；
```

图 8-55　事件类

}

这样我们在处理相应的 Event 的时候,就可以通过接口中的 getEvents 方法直接返回,而不用理会返回的是哪个事件列表里面的 Event。

(3) 观察者模式。由于系统是基于事件驱动的,所以在用户的请求过程中,将会发送事件到事件列表进行处理。为了在处理完成后通知用户,我们在系统中引入了回调机制,每个事件都会绑定一个相应的观察者,在执行完事件后会执行相应的观察者绑定的方法(代码 8-6)。

【代码 8-6】 IEvent 接口
```
public interface IEvent extends Serializable{
    public void doThings();
    public void addResourceByName(String resourceName);
    public void notifyResource();
}
```

IEvent 接口通过 addResourceByName 方法将相应的 Resource 注册到事件中,当完成 doThings 后会通过 notifyResource 方法通知相应的 Resource。

8.3.3 领域层

1. 主要的类图

资源库(如图 8-56 所示)是由一个封装过的 ConcurrentMap 形成的 Key-Value 数据库,整个对象都保存在内存中,里面的模型实体保存着数据。IResource 接口定义了对 Model 的各种操作方法。由于系统是并发访问的,所以得保证数据的同步,这样方法都是封装了 Map 的原子操作方法。

图 8-56 资源库接口

模型类（如图8-57和8-58所示）是系统中的实体，主要是保存着信息和对其他实体的引用。模型都实现了Serializable，这样可以保证序列化。

图8-57 模型类

Domain下面的服务类（如图8-59所示）是返回多个实体组成的数据的方法。

2. 方法思想

（1）领域驱动设计（Domain Driven Design）。领域驱动设计是一种软件开发方法，目的是在实现软件系统时，能准确地对真实的业务过程建模，并可根据真实的业务过程的调整而调整。

传统的开发工作趋向于一种以技术为先导的过程，需求从业务方传递到开发团队，开发人员依据需求上的描述创造出最有可能的假想。

随着软件系统的开发和发展，你对各种问题的理解也会更深刻。领域驱动设计就是要通过深入地理解问题来找到问题的解决方案。

领域驱动设计真正的不同之处是，它可以把软件系统当作业务过程的一个映射，是使之能动，而不是驱动之。领域驱动设计是要你深入到业务过程中，了解业务术语和实践方法。技术方面的事被放在了第二位。

DDD最大的好处是：接触到需求后的第一步是考虑领域模型，而不是将其切割成

图 8-58　模型类

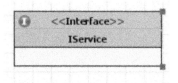

图 8-59　服务类

数据和行为，然后数据用数据库实现，行为使用服务实现，最后造成需求的首肢分离。DDD 让你首先考虑的是业务语言，而不是数据。考虑的重点不同会导致编程世界观不同。

　　DDD 划分聚合（如图 8-60 所示）的依据是根据实体之间的关系。每个聚合将会由一个 Resource 来进行管理。和其他聚合有关系的称为聚合根，聚合根会保持对其他实体对象的引用，通过聚合根的相应操作可以实现对所有实体的操作。

　　因为 Oe_Section 和 Oe_Course 只与实体 Oe_Class 有关系，所有可以将这三个实体划分成一个聚合（如图 8-61 所示），其聚合根为 Oe_Class。因此，我们可以为这个聚合分配一个 Repository。

图 8-60 聚合

图 8-61 聚合一

【代码 8-7】聚合类 ClassRepository

```
public class ClassRepository extends AResource<Integer, OeClass>{
    private static ClassRepository classRepository = new ClassRepository();
    ...
}
```

ClassRepository（代码 8-7）继承自 AResource 并且持有自己的静态实例，Oe_Section 和 Oe_Course 的实例则通过 Oe_Class 的方法返回。

Oe_Record 聚合、Oe_Status 实体只和 Oe_Record 实体有关，所以这两个实体为一个聚合（如图 8-62 所示）。其中 Oe_Record 是聚合根，通过 Oe_Record 可以找到 Oe_Status 的实例。同样，我们可以为这个聚合分配一个 Repository。

图 8-62 聚合二

【代码 8-8】聚合类 ClassRepository 2

```
public class RecordRepository extends AResource<UUID, OeRecord>{
    private static RecordRepository recordRepository = new RecordRepository();
    ...
}
```

RecordRepository（代码 8-8）继承自 AResource 并拥有自己的一个静态实例，Repository 保存着一个 Oe_Record 的 Map，Oe_Status 实例可以通过 Oe_Record 的方法获得。

Oe_User 聚合、Oe_Roles 实体只和 Oe_User 实体有关系，所有这三个实体可以划分成一个聚合（如图 8-63 所示），Oe_User 是聚合根。由 Oe_User 保存着 Oe_Roles 的实例。我们可以为这个聚合分配一个 Repository。

图 8-63 聚合三

【代码8-9】聚合类 ClassRepository 3

```java
public class UserRepository extends AResource<Integer, OeUser>{
    private static UserRepository repository = new UserRepository();
    ...
}
```

UserRepository（代码8-9）通过继承 AResource 并持有自己的一个静态实例。UserRepository 保存着 OeUser 的一个 Map。

（2）泛型。泛型是 Java SE 1.5 的新特性，泛型的本质是参数化类型，也就是说所操作的数据类型被指定为一个参数。使用泛型可以为开发带来方便。

IResource（代码8-10）是一个泛型接口，其中分别实例化 E 和 T 就可以。方法中将会将参数 E 和 T 替换成相应的实例。泛型 UserRepository（代码8-11）类似。

【代码8-10】IResource 接口

```java
public interface IResource<E, T> {
    public String getName();
    public void addModel(T model);
    public void delModel(T model);
    public void delModel(T model, T old);
    public void updateModel(T model);
    public void updateModel(T model, T old);
    public IModel getModel(E modelKey);
    public ConcurrentMap<E, T> getModels();
    public void setModels(Object object);
    public void addInWareHouse(WareHouse wareHouse);
}
```

【代码8-11】泛型 UserRepository

```java
public class UserRepository extends AResource<Integer, Oe_User>{
    private static UserRepository repository = new UserRepository();
    public UserRepository() {
        //TODO Auto-generated constructor stub
        initModels();
        initName();
        addInWareHouse(WareHouse.getWareHourse());
    }
    @Override
    public String getName() {
        //详细实现…
        return null;
    }
    @Override
    public void addModel(OeUser model) {
```

```java
        //详细实现…
    }
    @Override
    public void delModel(OeUser model) {
        //详细实现…
    }
    @Override
    public void delModel(OeUser model, OeUser old) {
        //详细实现…
    }
    @Override
    public void updateModel(OeUser model) {
        //详细实现…

    }
    @Override
    public void updateModel(OeUser model, OeUser old) {
        //详细实现…
    }
    @Override
    public IModel getModel(Integer modelKey) {
        //详细实现…
        return null;
    }
    @Override
    public ConcurrentMap<Integer, OeUser> getModels() {
        //详细实现…
        return null;
    }
    @Override
    public void setModels(Object object) {
        //详细实现…
    }
    @Override
    public void addInWareHouse(WareHouse wareHouse) {
        //TODO Auto-generated method stub
    }
    @Override
    protected void initModels() {
        //详细实现…
    }
    @Override
    protected void initName() {
```

```
        //详细实现…
    }
}
```

（3）内存 Key – Value 数据缓存。基于 In – memory 的思想，系统需要快速命中目标而不需要太多耗时的 I/O 操作，所以我们会把数据都保持在内存中，这样读写就只发生在内存，并不会出现 I/O 操作的过程，可以加快相应的速度，这样对内存数据缓存的要求是能够快速定位到元素。因为我们会存储大量的数据，如果在查找和插入所耗费的时间更多，那这样就失去了意义。它还必须能够自动扩充容量，如果容量不能自动扩充，这样很容易出现 OOM 错误。

快速定位元素属于算法和数据结构的范畴，哈希（Hash）算法是一种简单可行的算法。所谓哈希算法，是将任意长度的二进制值映射为固定长度的较小二进制值。

哈希算法将任意长度的二进制值映射为固定长度的较小二进制值，这个小的二进制值称为哈希值。哈希值是一段数据唯一且极其紧凑的数值表示形式。如果散列一段明文且只更改该段落的一个字母，随后的哈希都将产生不同的值。要找到散列为同一个值的两个不同的输入，在计算上是不可能的，所以数据的哈希值可以检验数据的完整性。

通过哈希算法，一个元素映射到了某个位置，现在一旦扩充了容量，也就意味着元素映射的位置需要变化。对于哈希算法来说，调整了映射的小集合，那么原来映射的路径肯定就不复存在，就需要重新计算映射路径，也就是所谓的 rehash 过程。

我们实现的数据缓存，是以 HashMap 为基础的数据缓存。在 HashMap 中首先由一个对象数组 table 组成，修饰符 transient 表示序列号的时候不被存储。size 描述的是 Map 中元素的大小，threshold 描述的是达到指定元素个数后需要扩容，loadFactor 是扩容因子（loadFactor > 0），也就是计算 threshold。元素的容量是 table.length，即数组的大小。换句话说，如果存取的元素大小达到了整个容量（table.length）的 loadFactor 倍（也就是 table.length * loadFactor），那么就需要扩充容量。在 HashMap 中，每次扩容都是扩大数组的一倍，使数组大小变为原来的两倍。

由于 HashMap 扩容的过程非常耗时，将会导致所有的数据重新进行 hash 计算，所以容器初始化大小必须设置得比较合理，避免 HashMap 多次扩容，而且容量越大，re-Hash 的过程所耗费的时间就越多。

由于数据缓存在内存中只有一个实例，所以就必须保证数据的同步，这里我们采用了 ConcurrentHashMap，默认情况下 ConcurrentHashMap 是用了 16 个类似 HashMap 的结构，其中每一个 HashMap 拥有一个独占锁。最终的效果就是通过某种 Hash 算法，将任何一个元素均匀地映射到某个 HashMap 的 Map.Entry 上面，而对某个一个元素的操作就集中在其分布的 HashMap 上，与其他 HashMap 无关。这样就支持最多 16 个并发的写操作。而且通过 ConcurrentHashMap 的原子方法可以实现线程安全。

数据缓存的实现都是封装自 ConcurrentHashMap 的同步的方法，这样就可以保持数据同步。

8.3.4 应用层

1. 主要的类

（1）系统初始数据（如图 8-64 所示）可以从硬盘序列化文件上面获取，该线程可以设置为自启动。启动时，会通过路径 Path 找到相应的序列化数据并将其反序列化，然后保存在内存中。

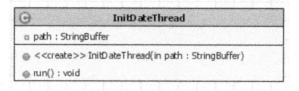

图 8-64　初始化数据线程

（2）将数据缓存中的数据序列化（如图 8-65 所示）到硬盘上，图 8-66 为工作线程。

图 8-65　序列化数据线程

图 8-66　工作线程

（3）执行 Event 的 doThings 方法，并且调用 Event 的 notifyResource 方法通知 Resource 对数据进行更新。

（4）TaskThread 线程池（如图 8-67 所示），通过 ThreadPoolExecutor 保存一定的线

程池,并在需要的时候添加更多的TaskThread。

图8-67 工作线程池

线程池可以解决两个不同问题:由于减少了每个任务调用的开销,它们增强在执行大量异步任务时的性能,并且可以提供绑定和管理资源(包括执行任务集时使用的线程)的方法。

使用ThreadFactory创建新线程。

图8-68 自启动任务

(5)通过实现ServletContextListener并且在web.xml中配置成监听器可以达到自启动(如图8-68所示)的效果,在启动时将会调用contextInitialized方法,启动一些配置上的线程。在销毁时会调用contextDestroyed方法,在销毁的时候停止所有线程(代码8-12)。

【代码8-12】listener

```
<listener>
    <listener-class>
        cn.nfsysu.csip.oe.application.AutoExecute
    </listener-class>
</listener>
```

2. 方法思想

优化存储。因为数据必须从内存中存储到外存,数据传递、持久化、I/O都是最耗时间的,所以我们使用内存数据缓存机制,将系统的数据保存在内存中,这样存取速度将大大提高。但是,保存在内存中的数据是没有持久化的,如果内存断电就会导致数据的丢失,所以我们又必须将内存中的数据保存到外存中,这包含两种方法:存储进数据库和序列化到硬盘。同时,必须设计存储的规则,否则将会导致有数据就进行存储或者长时间不存储,这样缓存就没有优势。因为持续的存储必定会导致容器被占用,其他线程没法使用，所以会降低响应速度；而如果是长时间又都不进行存储，又容易造成数据的丢失。

基于此，我们设计了一个存储的算法，当没有进行任何数据写入操作时，数据存储线程是挂起的，也就是不会执行任何操作；当有数据写入的时候将会进行计时，如果在一个时间段内没有继续的写入操作则视为已经完成业务，然后便会唤醒数据存储线程。

对于数据安全这方面还进行另外的设计。通过保存写入事件，可以在保存完数据后再将事件清除，如果服务器出现宕机，启动时就会检查是否存在没有及时保存的数据，有的话就重新执行相应的事件，保存完再自动删除事件。

8.3.5 用户接口层

1. 主要的类

图 8-69 至图 8-92 为用户接口层主要的类。

```
LoginServlet
-------------------------------------------------
◇ doGet(in req : HttpServletRequest,in resp : HttpServletResponse) : void
◇ doPost(in req : HttpServletRequest,in resp : HttpServletResponse) : void
```

图 8-69　登录

```
LogoutServlet
-------------------------------------------------
◇ doGet(in req : HttpServletRequest,in resp : HttpServletResponse) : void
◇ doPost(in req : HttpServletRequest,in resp : HttpServletResponse) : void
```

图 8-70　登出

```
AddCourseServlet
-------------------------------------------------
◇ doGet(in req : HttpServletRequest,in resp : HttpServletResponse) : void
◇ doPost(in req : HttpServletRequest,in resp : HttpServletResponse) : void
```

图 8-71　添加课程

```
DeleteCourseServlet
-------------------------------------------------
◇ doGet(in req : HttpServletRequest,in resp : HttpServletResponse) : void
◇ doPost(in req : HttpServletRequest,in resp : HttpServletResponse) : void
```

图 8-72　删除课程

```
GetCourseServlet
-------------------------------------------------
◇ doGet(in req : HttpServletRequest,in resp : HttpServletResponse) : void
◇ doPost(in req : HttpServletRequest,in resp : HttpServletResponse) : void
```

图 8-73　获得课程

图8-74 修改课程

图8-75 添加用户

图8-76 批量添加用户

图8-77 删除用户

图8-78 获得用户

图8-79 获得用户列表

图8-80 修改用户

图 8-81 添加上课时间

图 8-82 删除上课时间

图 8-83 获得上课时间

图 8-84 修改上课时间

图 8-85 添加单元

图 8-86 删除单元

图 8-87 获得单元

图 8 – 88　修改单元

图 8 – 89　添加记录

图 8 – 90　获得班级记录

图 8 – 91　获得用户记录

图 8 – 92　修改记录

2. 方法思想

（1）不重复抽取（如代码 8 – 13 所示）。系统需要实现一个可以随机抽取名单的功能，这样既能减少考勤所使用的时间，还可以保证公平性。具体的算法思想是：假设数组长度为 n，每次从一个数组中随机抽取一个数字，数字的范围从 0 到 $n-1$，抽取完成后，将命中的元素赋值为数组的第 $n-1$ 个元素，然后再将 n 进行减 1 运算，这样循环进行运算，直到算法停止。该算法是扑克牌发牌算法的简化版，因为我们只需要对一个用户，也就是教师发牌，可以将目标容器减少到只有一个。

【代码 8 – 13】不重复抽取的例子

```
...
int n = oeUsers.length;
for(int i = 0; i < oeUsers.length; i++){
```

```
        int index = (int)(Math.random() * n);
        //目标值
        oeUser = oeUsers[index];
        //覆盖掉被选中的值,并且将指标减少1
        oeUsers[index] = oeUsers[n -1];
        n - -;
    }
    ...
```

（2）回溯思想。回溯法采用试错的思想，它尝试分步地解决一个问题。在分步解决问题的过程中，当它通过尝试发现现有的分步答案不能得到有效、正确的解答的时候，它将取消上一步甚至是上几步的计算，再通过其他可能的分步解答再次尝试寻找该问题的答案。

有时我们要得到问题的解，先从其中某一种情况进行试探，在试探过程中，一旦发现原来的选择是错误的，就退回一步重新选择。

在关于 unit 的设计中，我们使用了一个无穷层级表，也就是一棵树，如图 8 - 93 所示。

图 8 - 93 单元 unit 设计

图 8 - 93 中，以 unid 等于 1 的元素为根节点，层次为 0。2 和 4 都是 1 节点的子节点，而 3 和 5 分别是 2 和 4 的子节点，这是一个无穷层级的表，在实际应用上获取数据可能会比较麻烦，因为父节点和子节点都是同一个表，搜索起来比较耗费资源。我们可能会通过以父节点请求子节点的做法，但这样做的弊端就是会多出对服务器的请求，加大服务器的压力。这样会导致比较大的性能上的问题，所以我们必须在客户端进行处理。我们将数据处理成树的结果，然后对树进行遍历，这里采用的是深度优先算法，如代码 8 - 14。

【代码 8 - 14】深度优先算法

```
//深度优先算法
function GetNodes(result) {
    for (var i = 0; i < result.length; i + +) {
        html + = " <li > " + result[i].name;
        if (result[i].childs ! = undefined) {
            html + = " <ul > ";
            GetNodes(result[i].childs);
```

```javascript
        html += "</ul>";
    }
    html += "</li>";
    }
    return html;
}
//setting 的格式:[ID,Name,PID]
//id 子节点,pid 父节点
function ToForest(sNodes, setting) {
    var i, l,
    //主键 ID
    key = setting[0];
    //parentID
    parentKey = setting[1];
    //childs
    childsKey = "childs";
    //参数检查
    if (!key || key == "" || !sNodes)
        return [];
    if ($.isArray(sNodes)) {
        //存放树形式的数据模型
        var r = [];
        //存放以 ID 为 key,ID 对应的实体为 value
        var tmpMap = [];
        //赋值操作
        for (i = 0; i < sNodes.length; i++) {
            //获取当前的 id
            var id = sNodes[i][key];
            tmpMap[id] = sNodes[i];
        }
            //对 json 逐层遍历确定层级关系
        for (i = 0; i < sNodes.length; i++) {
            //获取当前的 pid
            var pid = sNodes[i][parentKey];
            //判断是否是顶级节点
            if (tmpMap[pid]) {
                //判断该节点是否有孩子节点
                if (!tmpMap[pid][childsKey])
                    tmpMap[pid][childsKey] = [];
                //将此节点放在该节点的孩子中
                tmpMap[pid][childsKey].push(sNodes[i]);
            }
```

```
        else{
    //如果是顶级节点直接存放
    r.push(sNodes[i]);
    }
  }
    return r;
  }
    else{
    return [sNodes];
    }
}
```

（3）Java 远程方法调用。Java 远程方法调用，即 Java RMI（Java Remote Method Invocation）是 Java 编程语言里，一种用于实现远程过程调用的应用程序编程接口，它使客户机上运行的程序调用远程服务器上的对象。该方法使 Java 编程人员能够在网络环境中分布操作。RMI 全部的宗旨就是尽可能地简化远程接口对象的使用。

Java RMI 极大地依赖于接口。在需要创建一个远程对象的时候，程序员通过传递一个接口来隐藏底层的实现细节。客户端得到的远程对象句柄正好与本地的根代码连接，由后者负责通过网络通信。这样一来，程序员只需关心如何通过自己的接口句柄发送消息即可。

接口的两种常见实现方式是：最初使用 JRMP（Java Remote Message Protocol，Java 远程消息交换协议）实现；还可以用与 CORBA 兼容的方法实现。RMI 一般指的是编程接口，也有时候同时包括 JRMP 和 API（应用程序编程接口），而 RMI – IIOP 则一般指 RMI 接口接管绝大部分的功能，以支持 CORBA 的实现。

最初的 RMI API 设计为通用地支持不同形式的接口实现。后来，CORBA 增加了传值（pass by value）功能，以实现 RMI 接口。然而 RMI – IIOP 和 JRMP 实现的接口并不完全一致。

该方法所使用的 Java 包的名字是 java.rmi。

RMI 指的是远程方法调用（Remote Method Invocation）。它是一种机制，能够让在某个 Java 虚拟机上的对象调用另一个 Java 虚拟机中的对象上的方法。用此方法调用的任何对象必须实现该远程接口。调用这样一个对象时，其参数为"marshalled"，并将其从本地虚拟机发送到远程虚拟机（该远程虚拟机的参数为"unmarshalled"）上。该方法终止时，将获取来自远程机的结果并将结果发送到调用方的虚拟机。如果方法调用导致抛出异常，则将该异常指示给调用方。

该方法使用到的接口说明：

（1）Remote 接口用于标识可以从非本地虚拟机上调用的接口。任何远程对象都必须直接或间接实现此接口。只有在"远程接口"（扩展 java.rmi.Remote 的接口）中指定的这些方法才可远程使用。

UnicastRemoteObject 用于导出带 JRMP 的远程对象和获得与该远程对象通信的 stub。实现类可以实现任意数量的远程接口，并且可以扩展其他远程实现类。RMI 提供一些远程对象实现可以扩展的有用类,这些类便于远程对象创建,他们是 java.rmi.server.UnicastRemoteObject 和 java.rmi.activation.Activatable。代码 8-15 为 RMI 端口设置代码,例程代码 8-16 为 UserChangeInterface 类。

【代码 8-15】 RMI 端口设置

```
UserChangeInterfaceImpl changeInterfaceImpl = new UserChangeInterfaceImpl();
                UserChangeInterface changeInterface = (UserChangeInterface)UnicastRemoteObject.exportObject(changeInterfaceImpl,0);
Registry registry = LocateRegistry.createRegistry(2001); //监听 2001 端口
registry.rebind("UserChangeInterface",changeInterface); //注册远程接口
```

【代码 8-16】 UserChangeInterface 类

```
public interface UserChangeInterface extends Remote{
    public OeUser addUser() throws RemoteException;
    public List<OeUser> addUserList() throws RemoteException;
    public OeUser delUser() throws RemoteException;
    public OeUser changeUser() throws RemoteException;
}
```

这样当用户信息被修改时,其他系统可以通过调用远程的接口来返回相应的用户信息并进行更新,从而实现多子系统信息共享。

8.3.6 数据库设计

数据库设计主要是如何建立应用系统的数据库,是信息系统开发和建设中的核心技术。为了支持相关程序运行,同时由于数据库应用系统的复杂性,数据库设计就变得异常复杂,因此最佳设计不可能一蹴而就,而只能是一种"反复探寻,逐步求精"的过程,即规划和结构化数据库中的数据对象以及这些数据对象之间关系的过程。

1. 数据库面向对象模型

Web 考勤系统中的主要对象有课程、用户、角色、考勤记录、节和单元等,各对象之间的关系如图 8-94 所示。

在该模型中,各模块的含义及联系如下:

(1) oe_class:上课模型,代表的是每周的课程安排,按周重复,如果是临时课程,则在周的位置设置相应的数值。

(2) oe_course:课程模型,保存课程的名称。

(3) oe_record:考勤记录模型,保存某个学生对于某节课的记录。

(4) oe_roles:角色模型,将权限作为一个对象。

(5) oe_section:节模型,保存每天的节数时间安排。

(6) oe_status:考勤状态模型,自定义考勤状态,如已到、迟到、缺勤等。

(7) oe_unit:单元模型,通过一个对自身的引用来保存一个树形结构,可以添加无穷层的结构,适合组织比较庞大的系统,如学校分为学院、系、专业、班级,可以自由规划。

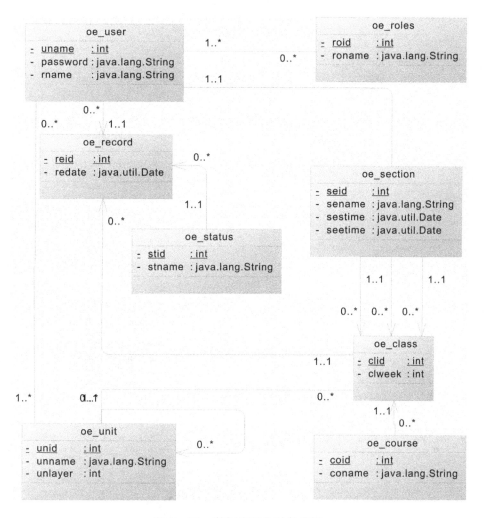

图 8-94 数据库面向对象模型

（8）oe_user：用户模型，保存用户的基本信息。

（9）用户和权限之间是多对多关系，一个用户拥有 1 到 N 个权限，权限不对用户持有引用。

（10）单元对自己持有一对多的关系，一个单元可以对应多个子单元。

（11）用户和记录是一对多关系，一个记录拥有一个用户。

（12）上课和记录是一对多关系，一个记录拥有一个上课。

（13）记录和状态是多对一关系，一个记录拥有一个状态。

（14）节和上课是一对多关系，一课拥有一个节。

（15）用户和上课是多对一的关系，一课拥有一个用户。

（16）课程和上课是一对多的关系，一课拥有一个课程。

2．数据库物理模型

本模型（如图 8-95 所示）采用了统一命名：简写 oe 加下画线再加模型名称。

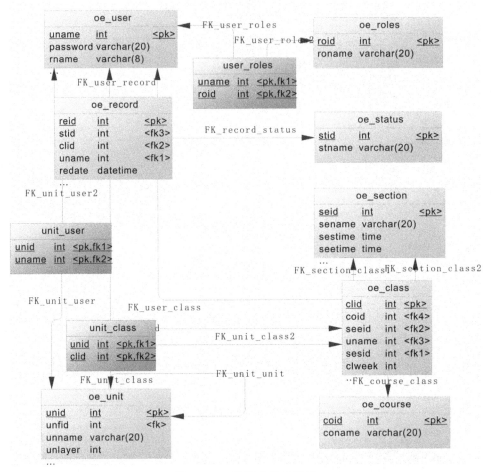

图 8-95 数据库物理模型

在该模型中,各模块的含义如下:

(1) oe_class:上课模型,代表的是每周的课程安排,以周重复,如果是临时课程,则在周的位置设置相应的数值。

(2) oe_course:课程模型,保存着课程的名称。

(3) oe_record:考勤记录模型,保存着某个学生对于某节课的记录。

(4) oe_roles:角色模型,将权限作为一个对象。

(5) oe_section:节模型,保存着每天的节数时间安排。

(6) oe_status:考勤状态模型,自定义考勤状态,如已到、迟到、缺勤等。

(7) oe_unit:单元模型,通过一个对自身的引用来保存一个树形结构,可以添加无穷层的结构,适合组织比较庞大的系统,如学校分为学院、系、专业、班级,可以自由规划。

(8) oe_user:用户模型,保存用户的基本信息。

3. 数据表设计和 SQL 脚本实现

表 8-2 上课表 oe_class

名称	类型	是否为空	主键	外键	说明
clid	int	not null	Pk		ID
coid	int	not null		Fk	课程 ID
seeid	int	not null		Fk	结束节 ID
uname	int	not null		Fk	用户 ID
sesid	int	not null		Fk	开始节 ID
clweek	int	not null			周几的课

【代码 8-17】

```
create table oe_class
(
clid     int not null,
coid     int not null,
seeid    int not null,
uname    int not null,
sesid    int not null,
clweek   int not null,
primary key (clid)
);
```

表 8-3 课程表 oe_course

名称	类型	是否为空	主键	外键	说明
coid	int	not null	Pk		ID
coname	varchar（20）	not null			课程名称

【代码 8-18】

```
create table oe_course
(
coid     int not null,
coname   varchar(20) not null,
primary key (coid)
);
```

表8-4 考勤记录表 oe_record

名称	类型	是否为空	主键	外键	说明
reid	int	not null	Pk		ID
stid	int	not null		Fk	状态ID
clid	int	not null		Fk	上课ID
uname	int	not null		Fk	用户ID
redate	datetime	not null			记录时间
releave	int				请假标志
recontent	varchar（50）				请假内容
reapprove	Int				批准请假

【代码8-19】
```
create table oe_record
(
reid      int not null,
stid      int not null,
clid      int not null,
uname     int not null,
redate datetime not null,
primary key (reid)
);
```

表8-5 角色表 oe_roles

名称	类型	是否为空	主键	外键	说明
roid	int	not null	Pk		ID
roname	varchar（20）	not null			权限名称

【代码8-20】
```
create table oe_roles
(
roid      int not null,
roname    varchar(20) not null,
primary key (roid)
);
```

表8-6 节表 oe_section

名称	类型	是否为空	主键	外键	说明
seid	int	not null	Pk		ID
sename	varchar（20）	not null			节名称
sestime	time	not null			开始时间
seetime	time	not null			结束时间

【代码8-21】
```
create table oe_section
(
seid       int not null,
sename     varchar(20) not null,
sestime    time not null,
seetime    time not null,
primary key (seid)
);
```

表8-7 考勤状态表 oe_status

名称	类型	是否为空	主键	外键	说明
stid	int	not null	Pk		ID
stname	varchar（20）	not null			状态名称

【代码8-22】
```
create table oe_status
(
stid       int not null,
stname     varchar(20) not null,
primary key (stid)
);
```

表8-8 单元表 oe_unit

名称	类型	是否为空	主键	外键	说明
unid	int	not null	Pk		ID
unfid	int			Fk	父单元ID
unnamed	varchar（20）	not null			单元名称
unlayer	int	not null			单元层次

【代码 8-23】
```
create table oe_unit
(
unid        int not null,
unfid       int,
unname      varchar(20) not null,
unlayer     int not null,
primary key (unid)
);
```

表 8-9 用户表 oe_user

名称	类型	是否为空	主键	外键	说明
uname	int	not null	Pk		ID
unid	int	not null		Fk	单元 ID
password	varchar（20）	not null			密码
rname	varchar（8）	not null			真实姓名
usphoto	varchar（50）				相片
usphone	char（11）				手机
ussphone	varchar（6）				短号
ussex	char（1）				性别

【代码 8-24】
```
create table oe_user
(
uname       int not null,
unid        int not null,
password    varchar(20) not null,
rname       varchar(8) not null,
primary key (uname)
);
```

表 8-10 用户权限表 user_roles

名称	类型	是否为空	主键	外键	说明
uname	int	not null	Pk		用户 ID
roid	int	not null	Pk		权限 ID

【代码 8-25】
```
create table user_roles
```

```
(
uname      int not null,
roid       int not null,
primary key (uname, roid)
);
```

表 8-11 用户单元关联表 unit_user

名称	类型	是否为空	主键	外键	说明
uname	int	not null	Pk		用户 ID
unid	int	not null	Pk		单元 ID

【代码 8-26】
```
create table unit_user
(
Uname      int not null,
unid       int not null,
primary key (uname, unid)
);
```

表 8-12 课和单元关联表 class_unit

名称	类型	是否为空	主键	外键	说明
clid	int	not null	Pk		课 ID
unid	int	not null	Pk		单元 ID

【代码 8-27】
```
create table class_unit
(
clid       int not null,
unid       int not null,
primary key (clid, unid)
);
```

【代码 8-28】关联表的 MySQL 脚本
```
alter table oe_class add constraint FK_course_class foreign key (coid)
    references oe_course (coid) on delete restrict on update restrict;

alter table oe_class add constraint FK_section_class1 foreign key (sesid)
    references oe_section (seid) on delete restrict on update restrict;

alter table oe_class add constraint FK_section_class2 foreign key (seeid)
    references oe_section (seid) on delete restrict on update restrict;
```

```
alter table oe_class add constraint FK_user_class foreign key (uname)
    references oe_user (uname) on delete restrict on update restrict;

alter table oe_record add constraint FK_class_record foreign key (clid)
    references oe_class (clid) on delete restrict on update restrict;

alter table oe_record add constraint FK_record_status foreign key (stid)
    references oe_status (stid) on delete restrict on update restrict;

alter table oe_record add constraint FK_user_record foreign key (uname)
    references oe_user (uname) on delete restrict on update restrict;

alter table oe_unit add constraint FK_unit_unit foreign key (unfid)
    references oe_unit (unid) on delete restrict on update restrict;

alter table oe_user add constraint FK_unit_user foreign key (unid)
    references oe_unit (unid) on delete restrict on update restrict;

alter table user_roles add constraint FK_user_roles foreign key (uname)
    references oe_user (uname) on delete restrict on update restrict;

alter table user_roles add constraint FK_user_roles2 foreign key (roid)
    references oe_roles (roid) on delete restrict on update restrict;
```

8.4 系统实现

在本节，我们着重介绍 Web 考勤系统中主要功能的实现。

8.4.1 用户界面关键算法实现

在考勤功能中，我们需要实现一个可以随机抽取名单的功能，这样既能减少考勤所使用的时间，还可以保证公平性。具体的算法思想是：假设数组长度为 n，每次随机抽取一个数字，数字的范围从 0 到 $n-1$，然后抽取完成后，以此数字为下标访问数组对应的元素。此算法是扑克牌发牌算法的简化版，如代码 8-29 所示。

【代码 8-29】学生考勤名单生成算法

```
Function changeAttendanceName() {

    if(n > 0){
        var index = parseInt(Math.random() * n);
        user = studentsArry[index];
         $('#attendance_name').attr("data-userid",user.userName).html(user.realName);
```

```
            studentsArry[index] = studentsArry[n-1];
            n--;
    }else{
        $('#attendance_name').removeAttr("data-userid").html("无");
    }
}
```

在教师提交考勤数据时,我们可以使用 post 方法提交考勤数据(见代码 8-30)。

【代码 8-30】 提交考勤数据的 post 方法

```
Function sumbitAttendanceList(userid,statusid){
    alert(user.userName);
    $.post(
        "/OnlineAttendance/attendance/oneAttendance.do",
        {"userId":userid,"statusId":statusid},
        function(data){
            varobj = new Function('return '+data)();
            if(obj.result){
                createSuccessAlert(obj.message);
            }
        });
}
```

代码 8-31 实现了检查对话框中内容是否合法字符的功能。

【代码 8-31】

```
Function IsChar(s){
        var Number = "0123456789.abcdefghijklmnopqrstuvwxyz-'ABCDEFGHI-JKLMNOPQRSTUVWXYZ`~!@#$%^&*()_";
        for (i=0;i<s.length;i++){
        //Check that current character isn't whitespace.
            var c = s.charAt(i);
            if (Number.indexOf(c) == -1)
                return false;
    }
    return true;
}
```

代码 8-32 实现了创建警告信息窗的功能。

【代码 8-32】

```
Function createAlert(content,type){
    type = type || "";
    var date = new Date();
    $(document.createElement('div'))
    .addClass("alert" + type)
    .addClass('fade')
```

```
        .addClass('in')
        .addClass('fixed_alert')
        .addClass('is_used_alert')
        .html("<button type = "button" class = "close" data - dismiss = "alert" >
&times;</button>")
        .append("<p><strong>" + date.getHours() + ":" + date.getMinutes()
+ " - - " + "</strong>" + content + "</p>")
        .click(function(){
            $(this).alert('close');
        })
        .appendTo( $('#alert_cotent'))
        .delay(10000).fadeOut(500,function(){
            $(this).removeClass("is_used_alert");
            $(this).attr("style","");
            destroyAlert();
        });
}
```

代码 8 – 33 实现了几个不同功能的警告框。

【代码 8 – 33】
```
function createSuccessAlert(content) {
    createAlert(content,"alert - success");
}
function createErrorAlert(content) {
    createAlert(content,"alert - error");
}
function createInfoAlert(content) {
    createAlert(content,"alert - info");
}
```

代码 8 – 34 实现了生成删除警告信息框的功能。

【代码 8 – 34】
```
function destroyAlert() {

    $('#alert_cotent div.alert').each(function(){
        if(! $(this).hasClass("is_used_alert")){
            $(this).delay(1000).alert('close');
        }
    });
}
```

8.4.2 领域层实现

本节详细介绍资源库 IResource、上课类 OeClass 和用户类 OeUser 类的实现,其他实现请查看系统源代码。

1. 资源库接口 IResource 的实现

资源库 IResource（如图 8-96 所示）是由一个封装过的 ConcurrentMap 形成的 Key-Value 数据库，整个对象都保存在内存中，里面保存着带数据的模型实体。IResource 接口（代码 8-35）定义了对 Model 的各种操作方法。由于系统中多个任务并发访问资源库，为了保证数据的同步，资源库的方法都封装了 Map 的原子操作方法。

图 8-96 资源库接口 IResource

【代码 8-35】
```
public interface IResource<E,T> {
/** 获得资源库的名称 */
public String getName();
/** 添加 model */
public T addModel(T model);
/** 添加 model 列表 */
public boolean addModelList(Collection<T> models);
/** 添加现有的 model 键值对 */
public void addModels(ConcurrentMap<E,T> models);
/** 删除 model */
public T delModel(T model);
/** 删除 model,如果 model 不等于 old 则不删除 */
```

```java
public boolean delModel(T model, T old);
/** 更新 model */
public T updateModel(T model);
/** 更新 model,如果 model 不等于 old 则不更新 */
public boolean updateModel(T model, T old);
/** 根据 id 获取 model */
public T getModelbyID(E modelKey);
/** 获取 model */
public T getModel(T model);
/** 获得资源库 */
public ConcurrentMap<E, T> getModels();
/** 获得 id 列表 */
public Set<E> getIdList();
/** 获得 model 列表 */
public Collection<T> getModelList();
/** 设置资源库 */
public void setModels(Object object);
/** 将资源库添加到仓库中进行管理 */
public boolean addInWareHouse(WareHouse wareHouse);
/** 是否存在传入的 id */
public boolean hasID(E id);
/** 是否存在传入的 model */
public boolean hasModel(T model);
}
```

2. 模型类的实现 ——上课类 OeClass

模型类 OeClass（如图 8-97 所示）是系统的基础类，封装了对其他相关联的模型的操作。模型类是系统的实体，主要保存信息和对其他实体的引用，并且模型实现了 Serializable 接口，从而保证序列化（代码 8-36）。

【代码 8-36】

```java
public interface IModel extends Serializable {

    public boolean searchByKey(String key);

}
```

上课类 OeClass（代码 8-37）实现了针对与上课的各种操作，包括增减课程、修改课程信息、管理课程教师和学生、考勤记录等。

【代码 8-37】

```java
public class OeClass implements IModel {
    private static final long serialVersionUID = -1267215707705809643L;
    /** 课 id */
    private int clid;
```

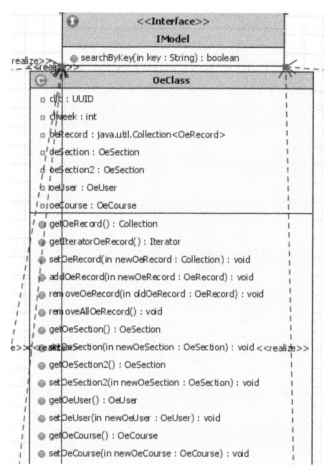

图 8-97 模型类 OeClass

```
/** 每周星期几 */
private int clweek;
/** 上课地点 */
private String clsite;
public int getClid() { return clid; }
public void setClid(int clid) { this.clid = clid; }
public int getClweek() { return clweek; }
public void setClweek(int clweek) { this.clweek = clweek; }
public String getClsite() { return clsite; }
public void setClsite(String clsite) { this.clsite = clsite; }
/** 该课的记录,一对多 */
private java.util.Collection<OeRecord> oeRecord;
/** 该课开课时间,多对一 */
private OeSection oeSection;
/** 该课下课时间,多对一 */
```

```java
    private OeSection oeSection2;
    /** 任课教师,多对一 */
    private OeUser oeUser;
    /** 课的名称,多对一 */
    private OeCourse oeCourse;
    /** 获得对应的班级,多对多 */
    private java.util.Collection<OeUnit> oeUnit;
    /** 获得该课的记录列表 */
    public java.util.Collection<OeRecord> getOeRecord() {
    if (oeRecord == null)
    oeRecord = new java.util.HashSet<OeRecord>();
    return oeRecord;
    }
    /** 获得记录列表的迭代项 */
    public java.util.Iterator<OeRecord> getIteratorOeRecord() {
    if (oeRecord == null)
            oeRecord = new java.util.HashSet<OeRecord>();
    return oeRecord.iterator();
    }
    /** 设置记录列表
    * @param newOeRecord
    */
    public void setOeRecord(java.util.Collection<OeRecord> newOeRecord) {
    removeAllOeRecord();
    for (java.util.Iterator<OeRecord> iter = newOeRecord.iterator(); iter.hasNext();)
    addOeRecord(iter.next());
    }
    /** 添加新纪录
    * @param newOeRecord
    */
    public void addOeRecord(OeRecord newOeRecord) {
    if (newOeRecord == null)
            return;
    if (this.oeRecord == null)
            this.oeRecord = new java.util.HashSet<OeRecord>();
    if (!this.oeRecord.contains(newOeRecord)) {
            this.oeRecord.add(newOeRecord);
            newOeRecord.setOeClass(this);
    }
    }
    /** 删除相应的记录
```

```java
 * @param oldOeRecord
 */
public void removeOeRecord(OeRecord oldOeRecord) {
    if (oldOeRecord == null)
        return;
    if (this.oeRecord != null) if (this.oeRecord.contains(oldOeRecord)) {
        this.oeRecord.remove(oldOeRecord);
        oldOeRecord.setOeClass((OeClass) null);
    }
}
/** 删除所有记录 */
public void removeAllOeRecord() {
    if (oeRecord != null) {
        OeRecord oldOeRecord;
        for (java.util.Iterator<OeRecord> iter = getIteratorOeRecord(); iter.hasNext();)
        {
            oldOeRecord = iter.next();
            iter.remove();
            oldOeRecord.setOeClass((OeClass) null);
        }
    }
}
/** 获得上课开始的节数 */
public OeSection getOeSection() {
        return oeSection;
}
/** 设置上课开始的节
 * @param newOeSection
 */
public void setOeSection(OeSection newOeSection) {
    if (this.oeSection == null || !this.oeSection.equals(newOeSection)) {
        if (newOeSection != null) {
            this.oeSection = newOeSection;
        }
    }
}
/** 获得下课的节数 */
public OeSection getOeSection2() {
        return oeSection2;
}
/** 设置下课的节数
```

```java
 * @param newOeSection
 */
public void setOeSection2(OeSection newOeSection) {
if (this.oeSection2 == null ||! this.oeSection2.equals(newOeSection)) {
    if (newOeSection != null) {
    this.oeSection2 = newOeSection;
    }
}
}
/** 获得该课的任课教师 */
public OeUser getOeUser() {
    return oeUser;
}
/** 设置该课的任课教师
 * @param newOeUser */
public void setOeUser(OeUser newOeUser) {
if (this.oeUser == null ||! this.oeUser.equals(newOeUser)) {
    if (this.oeUser != null) {
    OeUser oldOeUser = this.oeUser;
    this.oeUser = null;
    oldOeUser.removeOeClass(this);
    }
  if (newOeUser != null) {
    this.oeUser = newOeUser;
    this.oeUser.addOeClass(this);
    }
}
}
/** 获得该课的名称 */
public OeCourse getOeCourse() { return oeCourse; }
/** 设置该课的名称
 * @param newOeCourse
 */
public void setOeCourse(OeCourse newOeCourse) {
if (this.oeCourse == null ||! this.oeCourse.equals(newOeCourse)) {
    if (newOeCourse != null) {
    this.oeCourse = newOeCourse;
    }
}
}
/** 获得该课的单元列表 */
public java.util.Collection<OeUnit> getOeUnit() {
```

```java
if (oeUnit == null)
        oeUnit = new java.util.HashSet<OeUnit>();
return oeUnit;
}
/** 获得该课单元列表的迭代 */
public java.util.Iterator<OeUnit> getIteratorOeUnit() {
if (oeUnit == null)
        oeUnit = new java.util.HashSet<OeUnit>();
return oeUnit.iterator();
}
/** 设置该课的单元列表
 * @param newOeUnit
 */
public void setOeUnit(java.util.Collection<OeUnit> newOeUnit) {
removeAllOeUnit();
for (java.util.Iterator<OeUnit> iter = newOeUnit.iterator(); iter.hasNext();)
        addOeUnit(iter.next());
}
/** 添加课的单元
 * @param newOeUnit
 */
public void addOeUnit(OeUnit newOeUnit) {
if (newOeUnit == null)
        return;
if (this.oeUnit == null)
        this.oeUnit = new java.util.HashSet<OeUnit>();
if (!this.oeUnit.contains(newOeUnit)) {
        this.oeUnit.add(newOeUnit);
        newOeUnit.addOeClass(this);
}
}
/** 删除课的单元
 * @param oldOeUnit
 */
public void removeOeUnit(OeUnit oldOeUnit) {
if (oldOeUnit == null) return;
if (this.oeUnit != null) if (this.oeUnit.contains(oldOeUnit)) {
        this.oeUnit.remove(oldOeUnit);
        oldOeUnit.removeOeClass(this);
}
}
```

```java
/** 删除用户单元列表 */
public void removeAllOeUnit() {
if (oeUnit != null) {
    OeUnit oldOeUnit;
     for (java.util.Iterator<OeUnit> iter = getIteratorOeUnit(); iter.hasNext();) {
    oldOeUnit = iter.next();
    iter.remove();
    oldOeUnit.removeOeClass(this);
}
}
}
@Override
public boolean searchByKey(String key) {
//TODO Auto-generated method stub
return false;
  }
}
```

课程比较类 ClassComparator（代码 8-38）通过实现 Comparator 接口，根据 OeClass 里面的 OeSection 按时间进行排序，所以在进行比较时，会在类中获得 OeClass 类里面的 OeSection 对象，并且根据该对象的 getSestime 方法进行排序，这样我们在方法中得到的 OeClass 列表就可以是以 OeSection 为对象来排序的列表。

【代码 8-38】课程比较类 ClassComparator

```java
public class ClassComparator implements Comparator<OeClass> {
    @Override
    public int compare(OeClass o1, OeClass o2) {

    if (o1.getClweek() > o2.getClweek()) {
        return 1;
    }
    else if (o1.getClweek() < o2.getClweek()) {
        return -1;
    }
    else {
        String time1 = o1.getOeSection().getSestime();
        String time2 = o2.getOeSection().getSestime();
        int i = 3;
        i = time1.compareTo(time2);
        System.out.println(i);
        if (i <= 0) {
        return i;
        }
```

```
        else{
        return 1;
      }
    }
  }
}
```

我们根据 Collections. sort 就可以得到自定义排序的序列（代码 8 – 39）。

【代码 8 – 39】

…

//根据上课时间进行排序
Comparator < OeClass > classComparator = newClassComparator();
Collections.sort(oeClasses, classComparator);

…

3. 模型类的实现——用户类 OeUser

模型 OeUser 类（如图 8 – 98 所示）是系统的基础类，封装了对其他相关联的模型的操作。模型类是系统中的实体，主要是保存着信息和对其他实体的引用，并且模型通过实现 Serializable 接口，从而保证序列化。

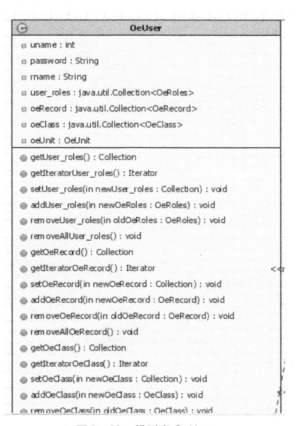

图 8 – 98 模型类 OeUser

OeUser类实现了用户的各种操作，包括对用户类别、用户权限、考勤记录等各种功能（代码8-40）。

【代码8-40】

```java
public class OeUser implements IModel{
private static final long serialVersionUID = -4825852053272948958L;
/** 用户id */
private String uName;
/** 用户密码 */
private java.lang.String passWord;
/** 用户名称 */
private java.lang.String rName;
/** 用户图片路径 */
private String usPath;
/** 用户手机 */
private String usPhone;
/** 用户短号 */
private String usSphone;
/** 用户性别 */
private char usSex;
public String getuName(){
    return uName;
}
public void setuName(String uName){
    this.uName = uName;
}
public java.lang.String getPassWord(){
    return passWord;
}
public void setPassWord(java.lang.String passWord){
    this.passWord = passWord;
}
public java.lang.String getrName(){
    return rName;
}
public void setrName(java.lang.String rName){
    this.rName = rName;
}
public String getUsPath(){
    return usPath;
}
public void setUsPath(String usPath){
```

```java
        this.usPath = usPath;
    }
    public String getUsPhone() {
        return usPhone;
    }
    public void setUsPhone(String usPhone) {
        this.usPhone = usPhone;
    }
    public String getUsSphone() {
        return usSphone;
    }
    public void setUsSphone(String usSphone) {
        this.usSphone = usSphone;
    }
    public char getUsSex() {
        return usSex;
    }
    public void setUsSex(char usSex) {
        this.usSex = usSex;
    }
    /** 用户角色列表,多对多 */
    private java.util.Collection<OeRoles> oeRoles;
    /** 用户记录列表,一对多 */
    private java.util.Collection<OeRecord> oeRecord;
    /** 用户任课列表,一对多 */
    private java.util.Collection<OeClass> oeClass;
    /** 用户对应单元列表,多 */
    private java.util.Collection<OeUnit> oeUnit;
    /** 获得用户角色列表 */
    public java.util.Collection<OeRoles> getOeRoles() {
        if (oeRoles == null)
            oeRoles = new java.util.HashSet<OeRoles>();
        return oeRoles;
    }
    /** 获得用户角色列表的迭代 */
    public java.util.Iterator<OeRoles> getIteratorOeRoles() {
        if (oeRoles == null)
            oeRoles = new java.util.HashSet<OeRoles>();
        return oeRoles.iterator();
    }
    /** 设置用户角色列表 *
     * @param newOeRoles
```

```java
     */
    public void setOeRoles(java.util.Collection<OeRoles> newOeRoles) {
        removeAllOeRoles();
        for (java.util.Iterator<OeRoles> iter = newOeRoles.iterator(); iter.hasNext();)
            addOeRoles(iter.next());
    }
    /** 添加用户角色
     * @param newOeRoles
     */
    public void addOeRoles(OeRoles newOeRoles) {
        if (newOeRoles == null) return;
        if (this.oeRoles == null)
            this.oeRoles = new java.util.HashSet<OeRoles>();
        if (!this.oeRoles.contains(newOeRoles)) {
            this.oeRoles.add(newOeRoles);
            //newOeRoles.addOeUsers(this);
        }
    }
    /** 删除用户角色
     * @param oldOeRoles
     */
    public void removeOeRoles(OeRoles oldOeRoles) {
        if (oldOeRoles == null) return;
        if (this.oeRoles != null)
            if (this.oeRoles.contains(oldOeRoles)) {
                this.oeRoles.remove(oldOeRoles);
            }
    }
    /** 删除用户角色列表 */
    public void removeAllOeRoles() {
        if (oeRoles != null) {
            for (java.util.Iterator<OeRoles> iter = getIteratorOeRoles(); iter.hasNext();) {
                iter.remove();
            }
        }
    }
    /** 获得用户记录列表 */
    public java.util.Collection<OeRecord> getOeRecord() {
        if (oeRecord == null)
            oeRecord = new java.util.HashSet<OeRecord>();
```

```java
            return oeRecord;
    }
    /** 获得用户记录列表的迭代 */
    public java.util.Iterator<OeRecord> getIteratorOeRecord() {
            if (oeRecord == null)
            oeRecord = new java.util.HashSet<OeRecord>();
            return oeRecord.iterator();
    }
    /** 设置用户记录列表
     * @param newOeRecord
     */
    public void setOeRecord(java.util.Collection<OeRecord> newOeRecord) {
            removeAllOeRecord();
            for (java.util.Iterator<OeRecord> iter = newOeRecord.iterator(); iter.hasNext();)
            addOeRecord(iter.next());
    }
    /** 添加用户记录
     * @param newOeRecord
     */
    public void addOeRecord(OeRecord newOeRecord) {
            if (newOeRecord == null)
            return;
            if (this.oeRecord == null)
            this.oeRecord = new java.util.HashSet<OeRecord>();
            if (!this.oeRecord.contains(newOeRecord)) {
            this.oeRecord.add(newOeRecord);
            newOeRecord.setOeUser(this);
            }
    }
    /** 删除用户记录
     * @param oldOeRecord */
    public void removeOeRecord(OeRecord oldOeRecord) {
            if (oldOeRecord == null)
            return;
            if (this.oeRecord != null)
            if (this.oeRecord.contains(oldOeRecord)) {
            this.oeRecord.remove(oldOeRecord);
            oldOeRecord.setOeUser((OeUser) null);
            }
    }
    /** 删除用户所有记录 */
```

```java
    public void removeAllOeRecord() {
        if (oeRecord != null) {
        OeRecord oldOeRecord;
    for (java.util.Iterator<OeRecord> iter = getIteratorOeRecord(); iter.hasNext();)
        {
        oldOeRecord = iter.next();
        iter.remove();
        oldOeRecord.setOeUser((OeUser) null);
        }
        }
    }
    /** 获得用户的任课列表 */
    public java.util.Collection<OeClass> getOeClass() {
        if (oeClass == null)
        oeClass = new java.util.HashSet<OeClass>();
        return oeClass;
    }
    /** 获得用户任课列表的迭代 */
    public java.util.Iterator<OeClass> getIteratorOeClass() {
        if (oeClass == null)
        oeClass = new java.util.HashSet<OeClass>();
        return oeClass.iterator();
    }
    /** 设置用户的任课列表
     * @param newOeClass */
    public void setOeClass(java.util.Collection<OeClass> newOeClass) {
        removeAllOeClass();
        for (java.util.Iterator<OeClass> iter = newOeClass.iterator(); iter.hasNext();)
        addOeClass(iter.next());
    }
    /** 添加用户任课
     * @param newOeClass
     */
    public void addOeClass(OeClass newOeClass) {
        if (newOeClass == null)
        return;
        if (this.oeClass == null)
        this.oeClass = new java.util.HashSet<OeClass>();
        if (!this.oeClass.contains(newOeClass)) {
        this.oeClass.add(newOeClass);
```

```java
        newOeClass.setOeUser(this);
        }
}
/** 删除用户任课
 * @param oldOeClass
 */
public void removeOeClass(OeClass oldOeClass) {
        if (oldOeClass == null)
        return;
        if (this.oeClass != null)
        if (this.oeClass.contains(oldOeClass)) {
        this.oeClass.remove(oldOeClass);
        oldOeClass.setOeUser((OeUser) null);
        }
}
/** 删除用户任课列表 */
public void removeAllOeClass() {
        if (oeClass != null) {
        OeClass oldOeClass;
        for (java.util.Iterator<OeClass> iter = getIteratorOeClass(); iter.hasNext();) {
        oldOeClass = iter.next();
        iter.remove();
        oldOeClass.setOeUser((OeUser) null);
        }
        }
}
/** 获得用户的单元列表 */
public java.util.Collection<OeUnit> getOeUnit() {
        if (oeUnit == null)
        oeUnit = new java.util.HashSet<OeUnit>();
        return oeUnit;
}
/** 获得用户单元列表的迭代 */
public java.util.Iterator<OeUnit> getIteratorOeUnit() {
        if (oeUnit == null)
        oeUnit = new java.util.HashSet<OeUnit>();
        return oeUnit.iterator();
}
/** 设置用户的单元列表
 * @param newOeUnit
 */
```

```java
public void setOeUnit(java.util.Collection<OeUnit> newOeUnit){
    removeAllOeUnit();
    for(java.util.Iterator<OeUnit> iter = newOeUnit.iterator();iter.hasNext();)
        addOeUnit(iter.next());
}
/** 添加用户单元
 * @param newOeUnit
 */
public void addOeUnit(OeUnit newOeUnit){
    if(newOeUnit==null)
        return;
    if(this.oeUnit==null)
        this.oeUnit = new java.util.HashSet<OeUnit>();
    if(!this.oeUnit.contains(newOeUnit)){
        this.oeUnit.add(newOeUnit);
        newOeUnit.addOeUser(this);
    }
}
/** 删除用户单元
 * @param oldOeUnit
 */
public void removeOeUnit(OeUnit oldOeUnit){
    if(oldOeUnit==null)
        return;
    if(this.oeUnit!=null)
        if(this.oeUnit.contains(oldOeUnit)){
            this.oeUnit.remove(oldOeUnit);
            oldOeUnit.removeOeUser(this);
        }
}
/** 删除用户单元列表 */
public void removeAllOeUnit(){
    if(oeUnit!=null){
        OeUnit oldOeUnit;
        for(java.util.Iterator<OeUnit> iter = getIteratorOeUnit();iter.hasNext();){
            oldOeUnit = iter.next();
            iter.remove();
            oldOeUnit.removeOeUser(this);
        }
    }
```

```
    }
    @Override
    public boolean searchByKey(String key){
        return false;
    }
}
```

用户模型的排序类：在系统的实现过程中，我们需要为自己的模型进行自定义排序，这时候 Java 提供的 unit 工具已经不能满足我们的需求，所有我们必须拓展 Java 提供的接口，为复杂的模型类提供自定义的排序。

UserComparator 类（代码 8 – 41）实现了 Comparator 接口，主要提供 OeUser 比较的方法，为 OeUser 模型类提供排序的功能。这里我们需要对用户里面的名称进行排序，通过对两个用户名的比较我们可以返回特定的标志，1 代表大于，也就是 "o1" 的名称的默认排序比 "o2" 的大，所以在该排序规则的影响下，我们获得的列表将会是以用户名为排序方式的列表，0 代表等于，-1 代表小于。

【代码 8 – 41】

```
public class UserComparator implements Comparator<OeUser>{

    @Override
    public int compare(OeUser o1, OeUser o2){
        int i = o1.getrName().compareTo(o2.getrName());
        if(i <= 0){
            return i;
        }
        else{
            return 1;
        }
    }
}
```

8.4.3 用户接口层实现

用户接口层提供了一个基类，该类通过封装领域层的方法，为应用层提供领域层中各个接口的调用方法。

1. 获取用户动作 API

代码 8 – 42 实现了 Web 界面上用于获取用户请求的（与 Domain 层的）接口，主要是获取用户的请求，例如：通过 "/log/login.do" 可以判断用户是进行 login 请求。

【代码 8 – 42】动作请求获取 API

```
/** 获得请求的动作
 *
 * @param request
 * @param response
```

```
 * @return
 */
protected String requestAction(HttpServletRequest request, HttpServletResponse response) {
    StringBuffer path = request.getRequestURL();
    String action = path.substring(path.lastIndexOf("/") + 1, path.lastIndexOf("."));
    return action;
}
```

2. 获取出勤状态列表 API

代码 8-43 实现了获取出勤状态列表的方法，应用层不需要进行多余的实例化操作，只需调用该接口即可获取完整的出勤状态列表。

【代码 8-43】获取出勤状态列表 API

```
/**
 * 获得状态列表
 * @param status 是否启用判断
 * @return
 */
protected Collection<OeStatus> getStatuss(char status) {
    Collection<OeStatus> oeStatuss = getStatuss();
    Iterator<OeStatus> iterator = oeStatuss.iterator();
    while (iterator.hasNext()) {
        OeStatus oeStatus = iterator.next();
        if (oeStatus.getStstatus() == status) {
            iterator.remove();
        }
    }
    return oeStatuss;
}
```

3. 发送字符串到前台的 API

代码 8-44 实现了把一个字符串发送到前台的编程接口（API）。

【代码 8-44】发送字符串到前台的 API

```
/**
 * 发送字符串数据到前台
 * @param response
 * @param jsonStr
 */
protected void printJson(ServletResponse response, String jsonStr) {
    try {
        PrintWriter out = response.getWriter();
        out.print(jsonStr);
```

```
            out.flush();
            out.close();
        } catch (IOException e) {
            e.printStackTrace();
        }
    }
}
```

4. 用户访问权限检验 API

代码 8-45 实现了两种判断用户是否具有访问权限的方法：第一种方法根据用户等级进行判断，第二种方法根据用户角色进行判断。这两种方法返回 true 时，表示用户有操作权限，返回 false 时表示用户没有操作权限。

【代码 8-45】 校验用户权限的两个编程接口

```
/**
 * 根据等级判断是否拥有相应等级的角色
 * @param level
 * @param request
 * @return
 */
protected Boolean hasPermission(int level, HttpServletRequest request) {
    OeUser oeUser = (OeUser) request.getSession().getAttribute(getUserSessionKey());
    for (OeRoles oeRoles : oeUser.getOeRoles()) {
        if (oeRoles.getRoright() <= level) {
            return true;
        }
    }
    return false;
}

/**
 * 根据名称判断是否拥有相应等级的角色
 * @param rolesName
 * @param request
 * @return
 */
protected Boolean hasPermission(String rolesName, HttpServletRequest request) {
    OeUser oeUser = (OeUser) request.getSession().getAttribute(getUserSessionKey());
    for (OeRoles oeRoles : oeUser.getOeRoles()) {
        if (oeRoles.getRoname().equals(rolesName)) {
            return true;
```

            }
        }
        return false;
}

5. 获取用户角色列表 API

代码 8-46 提供了获取用户权限角色列表的编程接口。

【代码 8-46】获取用户角色列表 API

```
/**
 * 获得用户相应的空间
 * @param oeUser
 * @return
 */
protected String getDomainUrl(OeUser oeUser) {
    int right = 99999;
    String permission = "";
    Iterator<OeRoles> rolesIter = oeUser.getIteratorOeRoles();
    while (rolesIter.hasNext()) {
        OeRoles oeRoles = rolesIter.next();
        if (oeRoles.getRoright() < right) {
            right = oeRoles.getRoright();
            permission = oeRoles.getRoname();
        }
    }
    return permission;
}
```

6. 字符串筛选 API

代码 8-47 对生成的字符串进行筛选，可以去掉不必要的数据或者陷入 OOM。

【代码 8-47】字符串筛选 API

```
/**
 * 设置 jsonlib 的过滤列表
 * @param prams
 * @return
 */
protected JsonConfig setJsonConfig(String... prams) {
    JsonConfig config = new JsonConfig();
    //只要设置这个数组,指定过滤哪些字段
    config.setExcludes(prams);
    return config;
}
```

7. 文件上传 API

代码 8-48 实现了文件上传 API。

【代码 8-48】

```java
protected String upLoadFile ( HttpServletRequest request, HttpServletResponse response,
String upLoadPath, String temp) throws Exception {
    //在解析请求之前先判断请求类型是否为文件上传类型
    booleanisMultipart = ServletFileUpload.isMultipartContent(request);
    if (isMultipart) {
        //文件上传处理工厂
        FileItemFactory factory = newDiskFileItemFactory();
        //创建临时文件目录
        File tempFile = newFile(request.getSession().getServletContext().getRealPath(temp));
        //设置缓存大小
        ((DiskFileItemFactory) factory).setSizeThreshold(1024 * 1024);
        //设置临时文件存放地点
        ((DiskFileItemFactory) factory).setRepository(tempFile);
        //创建文件上传处理器
        ServletFileUpload upload = newServletFileUpload(factory);
        //将页面请求传递信息最大值设置为50M
        upload.setSizeMax(1024 * 1024 * 50);
        //将单个上传文件信息最大值设置为6M
        upload.setFileSizeMax(1024 * 1024 * 6);
        //开始解析请求信息
        List items = null;
        items = upload.parseRequest(request);
        String fileName = null;
        //对所有请求信息进行判断
        Iteratoriter = items.iterator();
        while (iter.hasNext()) {
            FileItem item = (FileItem) iter.next();
            //信息为普通的格式
            if (item.isFormField()) {
            String fieldName = item.getFieldName();
            String value = item.getString();
            request.setAttribute(fieldName, value);
            }
            //信息为文件格式
            else {
                fileName = item.getName();
                int index = fileName.lastIndexOf("\");
                fileName = fileName.substring(index + 1);
                request.setAttribute("realFileName", fileName);
```

```
            //将文件写入
            String basePath = request.getSession().getServletContext()
.getRealPath(upLoadPath);
            File file = newFile(basePath, fileName);
            item.write(file);
        }
        request.getRequestDispatcher("/uploadsuccess.jsp").forward(request,
response);
        return fileName;
        }
        return null;
    }
}
```

8. 用户登录接口 LoginServlet

参照设计图8-99，代码8-49实现了用户登录接口 Login。用户登录接口通过用户名判断用户权限，并自动导向相应的页面。同时，对用户信息进行判断，如果正确，则发送相应的信息返回前台。

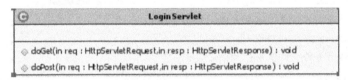

图8-99 登录接口

【代码8-49】

```
...
PrintWriter out = resp.getWriter();
OeUser checkUser = newOeUser();
checkUser.setuName(req.getParameter(this.getInitParameter("user_
name")));
checkUser.setPassWord(req.getParameter(this.getInitParameter("pass_
word")));
OeUser user = getUser(checkUser);
if(user!=null){
    //只有用户名
    if(checkUser.getPassWord()==null){
        if(user!=null&&user.getUsPath()!=null&&!user.getUsPath()
.equals("")){
            out.print(user.getUsPath());
        }else{
            out.print(MyConstants.ERROR);
        }
```

```
        } else {
          if (checkUser.getPassWord().equals(user.getPassWord())) {
              //将用户信息保存到 session
              req.getSession().setAttribute(getUserSessionKey(), user);
              //跳转到配置文件设置好的角色首页
            String url = "";
              System.out.println(getDomainUrl(user));
              if (getDomainUrl(user).equals("admin")) {
                  url = "/OnlineAttendance/admin/admin.html";
              } else if (getDomainUrl(user).equals("teacher")) {
                  url = "/OnlineAttendance/teacher/index.html";
              } else if (getDomainUrl(user).equals("student")) {
                  url = "/OnlineAttendance/student/index.html";
            }
              out.print(url);
          } else {
              out.print(MyConstants.ERROR);
          }
       }
        } else {
           out.print(MyConstants.ERROR);
       }
...
```

9. 用户登出接口 LogoutServlet

参照设计图 8 – 100，代码 8 – 50 实现了用户登出接口 Logout。用户登出时，清除页面缓存、清除 session、并且向前台发送信息，返回登录页面。

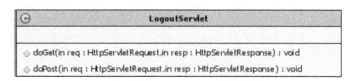

图 8 – 100　用户登出接口

【代码 8 – 50】
```
...
PrintWriterprintWriter = resp.getWriter();
//设置页面无缓存
resp.setHeader("Cache - Control", "no - cache, no - store");
resp.setHeader("Pragma", "no - cache");

req.getSession().invalidate();
printWriter.print(redirectURL);
```

...

10. 增加课程接口 AddCourseServlet

参照设计图 8-101，代码 8-51 实现了添加课程接口。添加课程接口，通过前台 post 请求获得课程基础数据并保持到系统中。

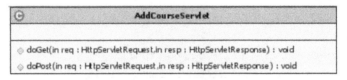

图 8-101 添加课程

【代码 8-51】

```
...
setBaseData(request, oeCourse);
isSuccess = (WEventBus.getEventBus().addEvent(
    newInsertEvent < OeCourse > (oeCourse, OeConstants.CourseRepository)) =
= null) ? true:false;
    if (isSuccess) {
        message = "添加课程成功!";
    } else {
        message = "添加课程失败!";
    }
...
```

11. 获取课程接口 GetCourseServlet

参照设计图 8-102，代码 8-52 实现获取课程的接口。获取课程接口，通过前台 post 请求获得相应课程。

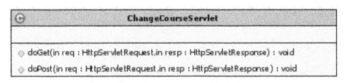

图 8-102 获得课程

【代码 8-52】

```
...
jsonArray = JSONArray.fromObject(getCourses());
jsonObject.put("courses", jsonArray);
...
```

12. 修改课程接口 ChangeCourseServlet

参照设计图 8-103，代码 8-53 实现修改课程接口。修改课程接口，通过前台 post 请求获得相应的数据并进行课程修改。

【代码 8-53】

...

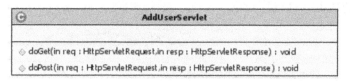

图 8-103 修改课程

```
oeCourse.setCoid(Integer.parseInt(request.getParameter("courseId")));
oeCourse = getCourse(oeCourse);
setBaseData(request, oeCourse);
isSuccess = true;
message = "修改课程成功!";
...
```

13. 增加用户接口 AddUserServlet

参照设计图 8-104，代码 8-54 实现增加用户接口。添加用户接口，通过前台 post 请求获得相应的数据并进行用户添加。

图 8-104 添加用户

【代码 8-54】

```
...
try {
    oeUser.setuName(request.getParameter("userId"));
    if (getUser(oeUser) == null) {
        isSuccess = addUser(request, response, oeUser);
    }
    System.out.println(isSuccess);
} catch (Exception e) {
    // TODO Auto-generated catch block
    e.printStackTrace();
}
if (isSuccess) {
    message = "添加用户成功!";
} else {
    message = "添加用户失败!";
}
...
```

14. 删除用户接口 DeleteUserServlet

参照设计图 8-105，DAIMA 8-55 实现删除用户接口。删除用户接口，通过前台

post 请求获得相应的数据并进行用户添加。

图 8-105 删除用户

【代码 8-55】

```
...
oeUser.setuName(request.getParameter("userName"));
    WEventBus.getEventBus().addEvent(new DeleteEvent<OeUser>(oeUser,
    OeConstants.UserRepository));
    ...
```

15. 获得用户接口 GetUserServlet

参照设计图 8-106，代码 8-56 实现获取用户接口。获得用户接口，通过前台 post 请求获得相应的数据并进行用户添加。

图 8-106 获得用户

【代码 8-56】

```
...
if(request.getParameter("userId")! = null&& ! request.getParameter("userId").equals("")){
        oeUser.setuName(request.getParameter("userId"));
        if(getUser(oeUser)! = null){
            isSuccess = false;
            message = "用户名已经存在";
        }else{
            isSuccess = true;
        }
    }
    ...
```

16. 获得用户列表接口 GetUserListServlet

参照设计图 8-107，代码 8-57 实现获取用户列表接口。获得用户列表接口，通过前台 post 请求获得相应的数据并进行用户添加。

【代码 8-57】

...

```
GetUserListServlet
```
　　doGet(in req : HttpServletRequest,in resp : HttpServletResponse) : void
　　doPost(in req : HttpServletRequest,in resp : HttpServletResponse) : void

图 8 – 107　获得用户列表

```
OeUnitoeUnit = new OeUnit();
oeUnit.setUnid(Integer.parseInt(request.getParameter("unitId")));
oeUnit = getUnit(oeUnit);
jsonArray = JSONArray.fromObject(unitUser(oeUnit));
jsonObject.put("unitUser", jsonArray);
jsonObject.put("unit", oeUnit.getUnname());
…
```

17. 修改用户接口 ChangeUserServlet

参照设计图 8 – 108，代码 8 – 58 实现修改用户接口。修改用户接口，通过前台 post 请求获得相应的数据并进行用户添加。

```
ChangeUserServlet
```
　　doGet(in req : HttpServletRequest,in resp : HttpServletResponse) : void
　　doPost(in req : HttpServletRequest,in resp : HttpServletResponse) : void

图 8 – 108　修改用户

【代码 8 – 58】
```
…
if (request.getParameter("userId") ! = null&& ! request.getParameter("use-
rId").equals("")) {
    if (request.getParameter("userOldId") ! = null
        && ! request.getParameter("userOldId").equals("")) {
        if (request.getParameter("userId").equals(
            request.getParameter("userOldId"))) {
            oeUser.setuName(request.getParameter("userId"));
            oeUser = getUser(oeUser);
            try {
                isSuccess = editUser(request, response, oeUser);
            } catch (Exception e) {
                //TODO Auto - generated catch block
                e.printStackTrace();
            }
        }
    }
    else {
        oeUser.setuName(request.getParameter("userOldId"));
```

```
                oeUser = getUser(oeUser);
                WEventBus.getEventBus().addEvent(
                                    newDeleteEvent<OeUser>(oeUser,
                                    OeConstants.UserRepository));
        oeUser.setuName(request.getParameter("userId"));
            try {
                isSuccess = addUser(request, response, oeUser);
            } catch (Exception e) {
                //TODO Auto-generated catch block
                e.printStackTrace();
            }
        }
    }
}
if (isSuccess) {
    message = "修改用户成功!";
}
else {
    message = "修改用户失败!";
}
...
```

18. 增加上课时间接口 AddClassServlet

参照设计图 8-109，代码 8-59 实现增加上课接口。

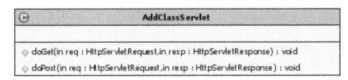

图 8-109 添加上课

【代码 8-59】
```
...
setClassInfo(request, response, oeClass);
WEventBus.getEventBus().addEvent(
newInsertEvent<OeClass>(oeClass, OeConstants.UserR-epository));
...
```

19. 删除上课时间接口 DeleteClassServlet

参照设计图 8-110，代码 8-60 实现删除上课时间接口。

【代码 8-60】
```
...
oeClass.setClid(Integer.parseInt(request.getParameter("")));
WEventBus.getEventBus().addEvent(
```

图 8-110 删除上课时间

```
                    newDeleteEvent < OeClass > ( oeClass,
OeConstants.UserRepository));
```
...

20. 获取上课时间接口 GetClassServlet

参照设计图 8-111，代码 8-61 实现获取上课接口。

图 8-111 获得上课时间

【代码 8-61】

```
...
    private void getCurriculum(ArrayList < OeClass > classes, JSONArrayjsonArray) {
        int flag = 0;
        booleanisOk = false;
        ArrayList < OeClass >oeClasses = classes;
        int size = oeClasses.size();
        Iterator < OeClass > iterator = oeClasses.iterator();
        OeClassoeClass = null;
        while (iterator.hasNext()) {
            oeClass = iterator.next();
            if (oeClass.getClweek() == DateUtil.getDayByWeek(new Date())) {
                intnowTime = (DateUtil.getHour() * 60 + DateUtil.getMinute());
                if ((Integer.parseInt(oeClass.getOeSection().getSestime()
                    .substring(0,2)) * 60 + Integer.parseInt(oeClass
                    .getOeSection().getSestime().substring(3))) >nowTime) {
                    isOk = true;
                }
                else {
                    isOk = false;
                }
            }
            else {
```

```
            isOk = false;
        }
        if (isOk) {
        break;
        }
        else {
            flag + +;
        }
}
for (int i = flag; i < size; i + +) {
        jsonArray.add(setCurriculum(oeClasses.get(i)));
}
        for (int i = 0; i < flag; i + +) {
        jsonArray.add(setCurriculum(oeClasses.get(i)));
    }
  }
...
```

21. 修改上课时间接口 ChangeClassServlet

参照设计图 8 – 112，代码 8 – 62 实现了修改上课时间接口。

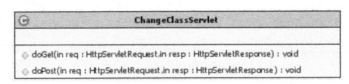

图 8 – 112 修改上课时间

【代码 8 – 62】
```
...
oeClass.setClid(Integer.parseInt(request.getParameter("")));
oeClass = getClass(oeClass);
setClassInfo(request, response, oeClass);
...
```

22. 添加单元接口 AddUnitServlet

参照设计图 8 – 113，代码 8 – 63 实现了增加单元接口。

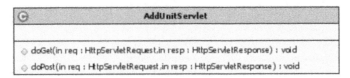

图 8 – 113 增加单元

【代码 8 – 63】
```
...
OeUnitoeUnit = new OeUnit();
JSONObjectjsonObject = new JSONObject();
booleanisSuccess = false;
String message = "";
if (requestAction(request, response).equals("addUnit")) {
    setUnit(request, response, oeUnit);
    isSuccess = (WEventBus.getEventBus().addEvent(
        newInsertEvent < OeUnit > (oeUnit, OeConstants.UnitRepository)) = =
null) ? true: false;
    if (isSuccess) {
        message = "添加单元成功!";
    }
    else {
        message = "添加单元失败!";
    }
}
...
```

23. 删除单元接口 DeleteUnitServlet

参照设计图 8 – 114，代码 8 – 64 实现删除单元接口。

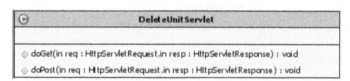

图 8 – 114　删除单元

【代码 8 – 64】
```
...
oeUnit.setUnid(Integer.parseInt(request.getParameter("")));
if (getUnit(oeUnit).getOeUser().size() = = 0) {
    //设置其他相关项,将被删除对象从其他对象的引用中去除
        WEventBus.getEventBus().addEvent(newDeleteEvent < OeUnit > (oeUnit,
    OeConstants.UnitRepository));
}
...
```

24. 获取单元接口 GetUnitServlet

参照设计图 8 – 115，代码 8 – 85 实现获取单元接口。

【代码 8 – 65】
```
private Collection < OeUnit > unitTree(OeUnitoeUnit, Stack < OeUnit > stack,
    Collection < OeUnit > oeUnits) {
```

```
                ┌─────────────────────────────────────────────────────┐
                │                  GetUnitServlet                      │
                ├─────────────────────────────────────────────────────┤
                │                                                     │
                ├─────────────────────────────────────────────────────┤
                │ ◆ doGet(in req : HttpServletRequest,in resp : HttpServletResponse) : void │
                │ ◆ doPost(in req : HttpServletRequest,in resp : HttpServletResponse) : void│
                └─────────────────────────────────────────────────────┘
```

图 8 – 115　获取单元接口

```
Stack < OeUnit > brotherStack = new Stack < OeUnit > ( );
    oeUnits.add(oeUnit);
    if (oeUnit.getOeUnitB( ).size( ) ! = 0) {
        for (OeUnit tmp : oeUnit.getOeUnitB( )) {
            brotherStack.push(tmp);
        }
        while (! brotherStack.empty( )) {
            stack.push(brotherStack.pop( ));
        }
    }
    if (! stack.empty( )) {
        unitTree(stack.pop( ), stack, oeUnits);
    }
    returnoeUnits;
}
```

25. 修改单元接口 ChangeUnitServlet

参照设计图 8 – 116，代码 8 – 66 实现修改单元接口。

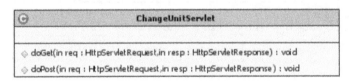

图 8 – 116　修改单元

【代码 8 – 66】
...
```
oeUnit.setUnid(Integer.parseInt(request.getParameter("unitId")));
oeUnit = getUnit(oeUnit);
setUnit(request, response, oeUnit);
isSuccess = true;
message = "修改单元成功!";
```
...

26. 增加考勤记录接口 AddRecordServlet

参照设计图 8 – 117，代码 8 – 67 实现增加考勤记录接口。
【代码 8 – 67】
...

图8-117 增加考勤记录接口

```
setRecord(request, response, oeRecord);
oeRecord.setReweek((short) DateUtil.getWeek());
WEventBus.getEventBus().addEvent(
                    newInsertEvent < OeRecord > (oeRecord,
OeConstants.RecordRepository));
...
```

27. 获取班级考勤记录接口 GetClassRecordServlet

参照设计图8-118，代码8-68实现获取班级考勤记录接口。

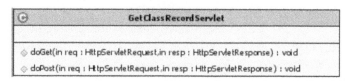

图8-118 获取班级考勤记录接口

【代码8-68】

```
...
if(hasPermission(4, request)) {
if(request.getParameter("classId") ! = null
        && ! request.getParameter("classId").equals("")) {
    OeClassoeClass = new OeClass();
      oeClass.setClid ( Integer.parseInt ( request.getParameter ( "clas-
sId")));
    oeClass = getClass(oeClass);
    int[] rate = rate(oeClass.getOeRecord());
    RateItem item = new RateItem();
    item.setCourse(oeClass.getOeCourse().getConame());
    item.setTotal(rate[0]);
    item.setNomal(rate[1]);
    item.setRate(rate[2]);
    item.setUserName(oeClass.getOeUser().getrName());
    jsonObject = JSONObject.fromObject(item);
  }
}
jsonArray = JSONArray.fromObject(recordBy(
    teacherRecord(oeUser, classId), week, status));
```

```
jsonObject.put("teacherRecord", jsonArray);
jsonArray = JSONArray.fromObject(getStatuss());
jsonObject.put("statuss", jsonArray);
...
```

28. 获取学生考勤记录接口 GetUserRecordServlet

参照设计图 8-119，代码 8-69 实现获取学生考勤记录接口。

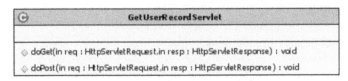

图 8-119　获取学生考勤记录接口

【代码 8-69】
```
...
jsonArray = JSONArray.fromObject(recordBy(
                studentRecord(oeUser, classId), week, status, 'a', 'a'));
jsonObject.put("studentRecord", jsonArray);
...
```

29. 修改考勤记录接口 ChangeRecordServlet

参照设计图 8-120，代码 8-70 实现修改学生考勤记录接口。

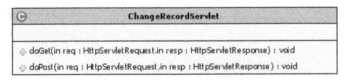

图 8-120　修改学生考勤记录接口

【代码 8-70】
```
...
oeRecord.setReid(Integer.parseInt(request.getParameter("recordId")));
oeRecord = getRecord(oeRecord);
setRecord(request, response, oeRecord);
...
```

8.4.4　应用层实现

应用层的功能是利用接口（User Interface）层提供的编程接口，实现 Web 考勤系统服务器。由于这部分主要是 JSP 代码的设计，本书不做详述，具体实现部分用户可以查看和运行 Web 考勤系统。

8.5 系统使用说明

8.5.1 运行环境

(1) 操作系统：X86/X64 位操作系统。
(2) JDK 的版本：JDK 1.7。
(3) 开发 IDE 环境：Eclipse 3.7。
(4) Web 服务器：Tomcat 7。
(5) 数据库：MySQL 5.0。

8.5.2 单应用服务器部署

单服务器具有简单配置、方便使用的特点，可以省去各种节点的配置。但是系统进行了数据共享，所以还需要对应用服务器进行配置。

项目中使用了 memcached-session-manager 库，这是一个使用 memcached 作为 tomcat session manager 的开源项目，在部署大型集群的时候，tomcat 自带的 session replication 技术会影响效率，使用统一的 session 存放策略更有利于集群规模的扩展，替换 session manager 的方法可以在程序代码不做修改的情况下实现。

将项目包放在 $TOMCAT_HOME/lib 目录下，然后修改配置文件 $TOMCAT_HOME/conf/server.xml，参见代码 8-71。

【代码 8-71】Tomcat 的配置文件 server.xml

```
< Context docBase = "E:/projects/test_msm/WebRoot" path = "" reloadable = "true" >
    < Manager className = "de.javakaffee.web.msm.MemcachedBackupSessionManager"
    memcachedNodes = "n1:localhost:11211"
    requestUriIgnorePattern = ".*\.(png|gif|jpg|css|js)$"
    sessionBackupAsync = "false"
    sessionBackupTimeout = "100"
    transcoderFactoryClass = "de.javakaffee.web.msm.serializer.javolution.JavolutionTranscoderFactory"
    copyCollectionsForSerialization = "false"
    />
```

在 tomcat 中添加 memcached 的节点，设置为 "n1：localhost：11211"，多台节点可以设置为："n1：localhost：11211 n2：localhost：11212"，即节点间使用空格分开。

这样可以实现 tomcat 的 session 共享，通过代码 8-72 可以在多应用中获得共享 session 的内容。

【代码 8-72】Tomcat 中 Session 共享的设置

```
session.setAttribute(new StringBuffer(oeUser.getUname()).toString(), oeUser);
```

```
oeUser = (OeUser) session.getAttribute(new StringBuffer(oeUser.getUname
()).toString());
```

在以上系统配置并部署成功后,可以运行 Web 考勤系统。基于本书篇幅,Web 考勤的详细例程可查看本书的教学资源包。

8.5 本章小结

本章设计并实现了一个基于 Web 的考勤系统,在章节安排上首先对考勤系统进行需求分析,确定系统的目标;然后进行界面设计、系统结构设计、系统实现,按照设计的增量分阶段实现每个增量要求的目标。

本章没有介绍 Java Web 项目开发所需技术,有兴趣的同学可查看其他相关教材。

第 9 章 基于 Android 平台的视频播放器

Android 应用开发是目前移动领域中最为火热的一个方向。众所周知，Android 应用也是基于 Java 语言开发的。在前面几个 Java 项目的基础上，本章准备开发一个基于 Android 平台的视频播放器，进一步巩固 Java 语言的基本知识，并将其扩展到 Android 平台中。考虑到读者可能对 Android 了解不多，所以本章先对 Android 做一简单的介绍，然后再介绍基于 Android 平台的视频播放器的开发过程。

9.1 Android 简介

Android 一词的本义指"机器人"，同时 Android 也是 Google 于 2007 年 11 月 5 日宣布的基于 Linux 平台开源手机操作系统名称，该平台由操作系统、中间件、用户界面和应用软件组成。Android 是基于 Java 并运行在 Linux 内核上的操作系统，这个操作系统是轻量级的，但是功能很全面。而且在这个操作系统上，Google 公司已经内置了很多有用的软件，如打电话、发短信等，是首个为移动终端打造的真正开放和完整的移动系统。2008 年 9 月 22 日，美国运营商 T‐Mobile USA 在纽约正式发布第一款 Google 手机——T‐MobileG1。该款手机为宏达电制造，是世界上第一部使用 Android 操作系统的手机，支持 WCDMA/HSPA 网络，理论下载速率 7.2Mbps，并支持 Wi‐Fi。

Google 与开放手机联盟（Open Handset Alliance）合作开发了 Android，这个联盟由包括高通、英特尔、德州仪器、中国移动、三星、摩托罗拉、高通、宏达电和 T‐Mobile 在内的 30 多家技术和无线应用的领军企业组成。Google 通过与运营商、设备制造商、开发商和其他有关各方结成深层次的合作伙伴关系，希望借助建立标准化、开放式的移动电话软件平台，在移动产业内形成一个开放式的生态系统。

Android 系统自推出以来，就以明显的优势逐渐扩大自身的市场份额，正处于蓬勃发展的阶段。据美国某市场调研机构 2012 年发布的一份最新报告显示：2012 年，一季度在美国基于 Android 系统的智能手机的销售量已占据全美手机销售份额的 28%，而大名鼎鼎的 iPhone 手机，其市场份额紧追其后，占到 21% 的市场份额。至今，Android 和 iOS 之间的系统大战还在如火如荼地进行。全球手机总销量中，智能手机销量所占的比例也迅速增长。2011 年全球智能手机销量首次超过 PC 电脑的销量，标志着"计算设备移动化"时代的到来。根据 IDC 最新的报告显示，Android 占全球智能手机的份额仍在上升。Android 设备占据的全球市场份额从 2012 年的 69% 上升到 2013 年的 78.6%。作为对比，iOS 设备全球市场份额反而从 2012 年的 18.7% 下降到了 2013 年的 15.2%。这就确定了 Android 系统的市场占有比的地位。业内人士分析，在开源理念日渐成为人们的共识的情况下，随着 Android 系统相应软件的不断开发应用，选择 Android 系统手机或者无线终端设备的人会越来越多，其市场霸主的地位在更新更好的系统出现之前是不可动摇的。

中国是世界上最大的手机消费国。由于 3G、4G 业务的不断推广，对整个手机业起大了巨大的促进作用，当前国内手机市场也正在快速向智能手机推进，而 Android 系统无疑是最大的市场需求，已经是当前世界上最流行的移动设备操作系统。各大中小型手机制造商近两年都在引入 Android 工程师，开发基于 Android 系统的智能手机。

未来基于 Android 系统的应用软件将进入飞速发展的全新阶段。Android 系统的应用绝不仅局限于手机产业，近几年来其迅速扩张到相关领域，如平板电脑、智能电视、车载系统、电视 STB、智能电器、智能会议系统、智能相机等。相信在不久的将来，还将有更多采用 Android 系统的高科技设备进入我们的生活。这些设备将会产生大量的应用需求，给应用开发者带来大量的机会。

Android 平台基于 Linux 内核，但并不是标准的 Linux。因为 Google 为了让 Android 更适合移动手持设备，对 Linux 内核进行了各种优化和增强。要学习和掌握 Android，首先需要对 Android 的系统架构有个了解，下面将详细介绍 Android 的系统架构。

9.1.1 Android 系统架构

Android 的系统架构采用了分层架构的思想，如图 9-1 所示。从上层到底层共包括四层，分别是应用程序层（Applications）、应用程序框架层（Application Framework）、系统运行库层（Libraries 和 Android Runtime）和 Linux 内核（Linux Kernel）。

图 9-1 Android 系统架构

1. 应用程序层

该层提供一些核心应用程序包，如短信、电话、日历、电子邮件、浏览器和联系人管理等。同时，开发者可以利用 Java 语言设计和编写属于自己的应用程序，而这些程序与 Android 核心应用程序彼此平等、友好共处。

2. 应用程序框架层

该层是 Android 应用程序开发的基础，开发人员大部分情况是利用该层提供的 API（应用程序接口）进行开发。应用程序框架层包括活动管理器、窗口管理器、内容提供者、视图系统、包管理器、电话管理器、资源管理器、位置管理器、通知管理器等，如表 9-1 所示。在 Android 平台上，开发人员可以访问核心应用程序所使用的 API 框架。并且，任何一个应用程序都可以发布自身的功能模块，而其他应用程序则可以使用这些已发布的功能模块。基于这样的重用机制，用户就可以方便地替换平台本身的各种应用程序组件。

表 9-1 应用程序框架层提供的服务

名称	功能介绍
窗口管理服务（Window Manager）	管理应用程序中的窗口（Window）
活动管理服务（Activity Manager）	管理应用程序中 Activity 组件的生命周期，比如创建 Activity、销毁 Activity 等
视图系统（View System）	应用程序通过视图（View）组件来显示各种信息，视图组件有 TextView、ImageView、Button、ListView 等
内容提供者（Content Providers）	应用程序可以通过内容提供者访问另一个应用程序的数据，如拨号应用程序通过内容提供者访问通讯录中的联系人数据
包管理服务（Package Manager）	管理所有安装在 Android 移动设备（如手机）的应用程序，比如安装和卸载应用程序
通知管理服务（Notification Manager）	使应用程序可以在通知栏显示自定义的通知信息
资源管理服务（Resource Manager）	提供各种资源供程序访问，如字符串资源、图片资源等
其他服务	Telephone Manager：电话服务 Use Manager：Usb 服务 Location Manager：位置服务 Bluetooth Manager：蓝牙服务 ……

3. 系统运行库层

如图 9-1 所示，系统运行库层包括两部分：程序库（Libraries）和 Android 运行时库（Android Runtime），下面分别介绍。

（1）程序库。Android 包含一些 C/C++库。应用程序通过应用程序框架层访问这些库提供的服务。表 9-2 介绍了这些库的功能。

表 9-2 程序库介绍

名称	描述
Surface Manager	管理访问显示子系统和从多模块应用中无缝整合 2D 和 3D 的图形
Media Framework	该库提供对音视频文件进行编解码、播放和录制等功能
Sqlite	Android 系统内置的轻量型关系型数据库引擎
OpenGL｜ES	3D 图形程序接口，是一个功能强大、调用方便的底层图形库。Android 所用到的是它的一个子集 OpenGL ES（OpenGL for Embedded Systems（嵌入式系统）），就是 OpenGL 的嵌入式版本
FreeType	矢量字体和位图显示引擎
WebKit	Web 浏览器引擎
SGL	2D 图形引擎，Android 使用 skia 作为核心的图形引擎
SSL	Secure Socket Layer 的首字母缩写，负责处理安全通信相关技术，为网络通信提供安全机制和数据完整性的一种通信协议
libc	Google 为嵌入式设备打造的 C 系统函数库

（2）Android 运行时库。Android 运行时库包括核心库和 Dalvik 虚拟机。核心库提供了 Java 语言核心库的大部分功能。Dalvik 虚拟机和一般 Java 虚拟机（Java VM）不同，它执行的不是 Java 标准的字节码文件（后缀名为.class），而是后缀名为.dex 的文件。二者最大的区别在于 Java VM 是基于栈的虚拟机（Stack-based），而 Dalvik 是基于寄存器的虚拟机（Register-based）。后者最大的好处在于可以根据硬件实现更大的优化，这更适合移动设备的特点。这也是 Google 采用 Dalvik 虚拟机的原因之一。

4. Linux 内核

Android 内核是基于 Linux 2.6 内核的，它是一个增强内核版本，除了修改部分 Bug 外，它提供了用于支持 Android 平台的设备驱动，其核心驱动主要包括：

（1）Android Binder：基于 OpenBinder 框架，用于提供 Android 平台的进程间通信（Inter Process Communication，IPC）功能。每一个 Android 应用程序都处于各自的进程中，都拥有一个 Dalvik 虚拟机实例。这些应用程序之间的通信就需要借助 Android Binder 来进行。

（2）Android 电源管理（Power Manager）：一个基于标准 Linux 电源管理系统的轻量级电源管理驱动，针对手机等嵌入式设备做了优化。

（3）低内存管理器（Low Memory Killer）：比 Linux 的标准的 OOM（Out Of Memory）Killer 机制更加灵活，它可以根据需要杀死进程以释放需要的内存。

（4）匿名共享内存（Anonymous Shared Memory）：为进程间提供大块共享内存，同时为内核提供回收和管理这个内存的机制。

(5) Android 调试桥（Android Debug Bridge，ADB）：嵌入式设备的调试比较麻烦，为了便于调试，Google 设计了这个调试工具，使用 USB 作为连接方式。ADB 可以看做是链接 Android 设备和 PC 机的一套机制。

(6) Android Logger：一个轻量级的日志设备，用于抓取 Android 系统的各种日志。日志功能对程序开发与调试非常有用，开发者可以从日志中提取出有用的信息从而帮助问题的定位和解决。

(7) Android Alarm：提供了一个定时器，用于把设备从睡眠状态唤醒，同时它还提供了一个即使在设备睡眠时也会运行的时钟基准。

(8) USB Gadget 驱动：一个基于标准 Linux USB gadget 驱动框架的设备驱动，Android 的 USB 驱动是基于 gadget 框架的。

(9) Android Ram Console：为了提供调试功能，Android 允许将调试日志信息写入一个被称为 RAM Console 的设备里，它是一个基于 RAM 的缓冲区（Buffer）。

(10) Yaffs2 文件系统：Android 采用 Yaffs2 作为 MTD（Memory Technology Devices）nand flash 文件系统。Yaffs2 是一个快速稳定的应用于 NAND 和 NOR Flash 的跨平台的嵌入式设备文件系统，同其他 Flash 文件系统相比，Yaffs2 能使用更小的内存来保存其运行状态，因此它占用内存小。Yaffs2 的垃圾回收非常简单而且快速，因此能表现出更好的性能。Yaffs2 在大容量的 NAND Flash 上的性能表现尤为突出，非常适合大容量的 Flash 存储。

9.1.2 Android 应用程序组成

Android 应用由 Activity、Services、Content Providers、Broadcast Receivers、Intents、Layout 等组件构成，其中 Activity、Services、Content Providers、Broadcast Receivers 是 Android 的四大基本组件。不是每一个 Android 应用程序都需要所有的这些组件，某些时候，只需要这其中的几种组合。以下对这些组件作一个具体的说明。

1. Activity（活动）

应用程序的表示层，Activity 就是一个有生命周期的对象。应用程序的每个界面都是 Activity 类的子类。

Activity 是 Android 中最常用的组件，一般通过 View 来实现应用程序的用户界面，相当于一个屏幕界面，用户与程序的交互是通过该类实现的。

如图 9-2 所示，Activity 具有生命周期，其生命周期主要包含三个状态，各种状态之间的切换是通过各种回调方法实现的。

（1）运行状态：正在与用户进行交互的 Activity 即是处于运行状态的 Activity，也就是说屏幕最上面的界面所对应的 Activity 即是处于运行状态的 Activity。处于运行状态的 Activity 获得了焦点，可以接受用户的按键及触摸事件并作出相应的响应。Activity 在处于运行状态之前，其相应的方法 onCreate（）、onStart（）、onResume（）依次被调用［注意，onCreate（）方法只有在 Activity 被创建时才会被调用，当该 Activity 由暂停状态再次回到运行状态时 onCreate（）不会被调用，而 onStart（）、onResume（）会被调用，参考图 9-2］。

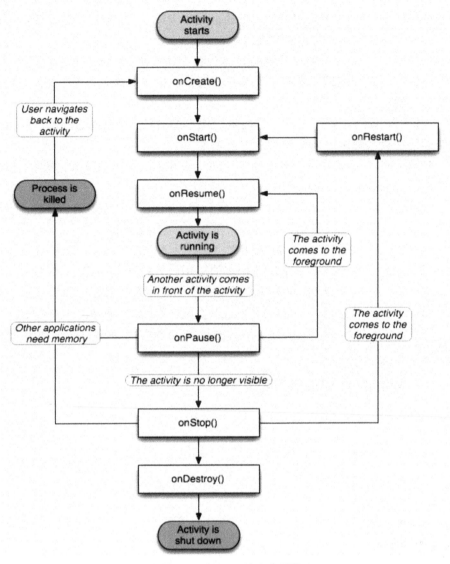

图 9-2　Activity 的生命周期

（2）暂停状态：当一个新的 Activity 启动后，之前处于活动状态的 Activity 将进入暂停状态，处于暂停状态的 Activity 失去焦点，新的 Activity 则获取到焦点并处于运行状态，当新的 Activity（前台显示的 Activity）不是全屏（如对话框形式的 Activity）时，可以见到暂停状态的 Activity。当一个 Activity 由运行状态进入暂停状态时，其 onPause（）方法将会被调用。当前台显示的 Activity 被销毁时，处于暂停状态的 Activity 有机会再次回到运行状态。

（3）停止状态：当一个 Activity 处于暂停状态并且完全不可见时，该 Activity 将进入停止状态。处于停止状态的 Activity 没有焦点，当系统内存不够用时，停止状态的 Activity 将被系统释放。当一个 Activity 由暂停状态进入停止状态时，其 onDestroy（）

方法将会被调用。

2. Service（服务）

Service（服务）它跟 Activity 的级别差不多，但不能自己运行并且只能在后台运行，可以和其他组件进行交互。Service 可以在很多场合的应用中使用，比如播放音乐的时候用户启动了其他 Activity，而这个时候程序要在后台继续播放音乐，这时就可以使用 Service 组件来实现；又比如检测 SD 卡上文件的变化，再或者在后台记录手机地理信息位置的改变等都可以使用 Service 组件来实现。总之服务总是藏在后台的，没有界面。Service 有两种启动方式，分别为调用 startService（）方法和 bindService（）方法来启动。两种启动方式 Service 的运行流程不一样。Service 跟 Activity 一样也具有生命周期，其生命周期的演变跟启动方式有关，如图 9-3 所示。

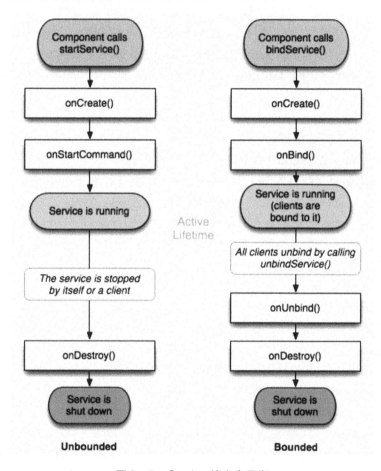

图 9-3　Service 的生命周期

3. Content Providers（内容提供者）

可共享的数据存储。Content Providers 用于管理和共享应用程序数据库。它是跨应用程序边界数据共享的优先方式。当应用程序中某个类继承了 Content Provider 类，并重写该类用于提供数据和存储数据的方法，就可以向其他应用共享其数据。虽然使用其他方法也可以对外共享数据，但数据访问方式会因数据存储的方式而不同，如采用文件方式对外共享数据，需要进行文件操作读写数据；采用 sharedpreferences 共享数据，需要使用 sharedpreferences API 读写数据。而使用 Content Provider 共享数据的好处是统一了数据访问方式。

4. Broadcast Receiver（广播接收器）

它是 Intent 广播的消费者（接收者）。如果应用程序创建并注册了一个 Broadcase Receiver，应用程序就可以监听匹配了特定过滤标准的广播 Intent。例如，当应用程序注册了接收一个开机启动完成后的 Broadcast Receiver，当手机启动完成后，系统会发送一个手机启动完成的广播，而应用程序则可以接收到该广播，从而得知系统已经启动完成。

5. Intent

中文翻译为"意图"，是应用程序基本组件间传递信息的一种数据载体。Intent 不但可以用于同一个应用程序中 Activities、Services 和 Broadcast Receivers 之间的交互，也可以用于不同应用程序之间 Activities、Services 和 Broadcast Receivers 的交互。

6. Layout

中文翻译为"布局"。Android 中有五大布局方式，分别是 LinearLayout（线性布局）、FrameLayout（框架布局）、AbsoluteLayout（绝对布局）、RelativeLayout（相对布局）和 TableLayout（表格布局），下面分别进行介绍。

（1）LinearLayout（线性布局）。LinearLayout 是项目开发中经常使用的布局，它能够根据用户为它设置的垂直或水平的属性值，在垂直方向或水平方向排列所有的子元素。由于所有的子元素都被堆放在其他子元素之后，因此垂直方向的线性布局的每一行只会有一个子元素，无论其多宽，水平方向的线性布局将会只有一个行高（高度为最高子元素的高度加上边框高度）。LinearLayout 保持子元素之间的间隔以及互相对齐（相对一个元素的右对齐、中间对齐或者左对齐）。

LinearLayout 还支持为单独的子元素指定 weight 属性值，这样做的好处是允许子元素可以填充屏幕上的剩余空间。这样也避免了在一个大屏幕中，一串小对象挤成一堆的情况，而是允许它们放大填充空白。为子元素指定一个 weight 值，剩余的空间就会按这些子元素指定的 weight 值按比例分配给这些子元素。默认的 weight 值为 0。例如，如果有三个文本框（TextView），其中两个指定了 weight 属性值为 1，那么，这两个文本框将等比例放大，并填满剩余的空间，而第三个文本框则不会放大。

（2）FrameLayout（框架布局）。FrameLayout 是最简单的一种布局方式，它被定制为屏幕上的一个空白备用区域，之后可以在其中填充一个单一对象，如一张图片。FrameLayout 的所有子元素将会固定在屏幕的左上角；不能单独为 FrameLayout 中的一个元素指定一个位置。后一个子元素将会直接在前一个子元素之上进行覆盖填充，前一个子元素部分或全部被后一个子元素挡住（除非后一个子元素是透明的）。

（3）AbsoluteLayout（绝对布局）。AbsoluteLayout 可以为子元素指定准确的 (x, y) 坐标值，并显示在屏幕上。$(0, 0)$ 为屏幕左上角，当向下移动时，y 坐标值将变大；而向右移动时，x 坐标值将变大。AbsoluteLayout 没有页边框，允许元素之间互相重叠

（尽管不推荐）。在项目开发中，通常不推荐使用 AbsoluteLayout，除非有足够的理由要使用它，因为它直接指定了元素相对于屏幕左上角的坐标，以至于在不同的设备上可能不能很好地工作。也就是说这种布局方式不能很好地做到屏幕适配，所以尽量避免使用这种布局方式。

（4）RelativeLayout（相对布局）。RelativeLayout 也是项目开发中经常使用的布局方式，它允许为子元素指定它们相对于其他元素或父元素的位置（通过元素的 ID 属性来指定）。子元素与子元素之间或子元素与父元素之间就有了相对关系。这样，可以以左右关系、上下关系或置于屏幕中央的方式来排列两个元素。由于元素按相对关系来排列，因此如果第一个元素位于屏幕的中央，那么相对于这个元素的其他元素将以屏幕中央为参照物来排列。

（5）TableLayout（表格布局）。TableLayout 布局将子元素的位置分配到行或列中。一个 TableLayout 由许多的 TableRow 组成，每个 TableRow 都会定义一个 row（一行）。TableLayout 容器不会显示 row、cloumns 或 cell 的边框线。每个 row 拥有零个或多个的 cell，每个 cell 拥有一个 View 对象。表格由列和行组成，包含许多的单元格，允许单元格为空，但单元格不能跨列，这与 HTML 中的不一样。

9.1.3　搭建 Android 开发环境

到此，读者应该对 Android 的基本框架和应用程序有个大概的了解。现在开始准备搭建 Android 开发环境。由于 Android 的应用是基于 Java 语言开发的，所以读者应先安装 JDK，安装 JDK 的具体步骤可以参考前面的章节。安装完 JDK 后，读者还需下载和安装 Android SDK 软件开发包。Google 为了方便初学者搭建开发环境，将开发 Android 应用所需的 Android SDK、ADT（Android Development Tools）以及 Eclipse 全部放在一个压缩文件（以 adt–bundle 为前缀文件名）中。读者只需下载对应的压缩文件即可。打开 IE 浏览器，在地址栏输入以下地址：http：//developer.android.com/sdk/installing/index.html？pkg = adt，如图 9 – 4 所示为该地址上的下载页面。

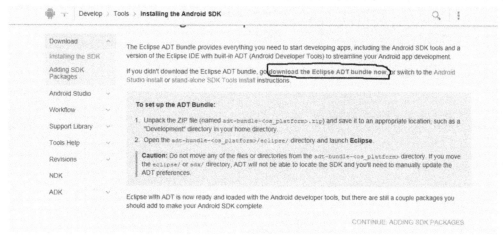

图 9 – 4　Android SDK 下载

单击超链接"download the Eclipse ADT bundle now"（图 9-4 中黑色框选中的部分），即可进入下载页面下载相应版本和平台的压缩文件。笔者下载的压缩文件为：adt-bundle-windows-x86-20130514.zip。解压该文件到相应的目录中，如 E：\ android 中，解压后的目录如图 9-5 所示。

图 9-5 解压后的文件目录

该目录有两个文件夹和一个 .exe 文件，其中 eclipse 文件夹放置了用于开发 Android 应用的集成开发环境 Eclipse，读者打开该文件夹，然后双击"eclipse.exe"即可启动 Eclipse 进行 Android 应用程序的开发；另一个文件夹为 sdk 文件夹，该文件夹中包含了开发 Android 应用所必需的 jar 文件、镜像文件（.img）以及一些开发工具如 aapt、ddms、adb、fastboot 等；SDK Manager.exe 可执行文件用来管理 SDK，负责 SDK 版本下载升级等功能。

至此，Android 开发环境就搭建完成了。接下来读者就可以开发自己的第一个 Android 应用了。

9.1.4 开发第一个 Android 应用程序

开发环境搭建完成，接下来就准备开发第一个 Android 应用。如同学习 C 语言和 Java 时一样，我们开发的第一个 Android 应用也是"Hello World"程序。如上一节所述，打开 Eclipse 文件夹，双击"eclipse.exe"启动 Eclipse。启动后的界面如图 9-6 所示（注意，笔者之前已经创建了几个 Android 应用，所以图 9-6 左边部分有一些项目显示，如果读者第一次打开 Eclipse，左边部分将是空白的）。

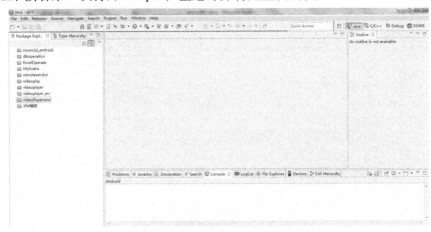

图 9-6 Eclipse 启动后的显示界面

启动 Eclipse 后，然后按照以下步骤创建 Android 项目。

（1）单击 File 菜单，选择"New"选项，在弹出的列表中选择"project…"选项，如图 9-7 所示。

图 9-7　创建 Hello World 项目（步骤 1）　　图 9-8　创建 Hello World 项目（步骤 2）

（2）选择"project…"选项后，会弹出一个对话框，选择"Android"目录下的"Android Application Project"，如图 9-8 所示。

（3）接下来弹出一对话框，要求输入 Application Name、Project Name 和 Package Name，在相应的文本框中输入名称，其他选项保持默认，如图 9-9 所示。

（4）单击【Next】按钮，勾选"create custom launcher icon"和"Create activity"复选框（默认为选中状态），如图 9-10 所示。

（5）单击【Next】按钮，出现如图 9-11 所示的对话框。

（6）单击【Next】按钮，出现如图 9-12 所示的对话框，选择"Blank Activity"。

（7）单击【Next】按钮，出现如图 9-13 所示的对话框，要求输入 Activity 和 Layout 文件的名字，保持默认不变。

（8）单击【Finish】按钮，完成 Hello World 工程的创建。这时，Eclipse 左边项目显示区将显示刚刚创建好的 Hello World 工程，如图 9-14 所示。

图 9-9　创建 Hello World 项目（步骤 3）　　图 9-10　创建 Hello World 项目（步骤 4）

图9-11 创建 Hello World 项目（步骤5）

图9-12 创建 Hello World 项目（步骤6）

图9-13 创建 Hello World 项目（步骤7）

图9-14 创建 Hello World 项目（步骤8）

HelloWorld 项目创建好后，该项目目录下生成很多文件夹及文件。如图9-14所示。下面对重点和常用的文件及文件夹进行介绍。

（1）src 文件目录为源文件目录，存放用户的项目源文件，所有可以被用户修改和创建的 Java 文件将被存放在这个目录下。

（2）gen 目录是自动生成的，用来保存 ADT 自动生成的文件，里面包含有 R.Java 文件，R.Java 文件用来存放全部资源文件的 ID，开发人员不需要修改该文件。

（3）assets 目录下存放应用程序中使用的外部资源文件，程序通过输入/输出流来对该目录中的文件进行读写操作，默认为空。

（4）bin 目录也是系统自动生成的目录，项目编译运行后，该目录下将会生成对应的 .dex 文件和 apk 文件。

（5）res 目录为项目的资源目录，里面又包括很多文件夹。其中，drawable-hdpi、drawable-mdpi、drawable-ldpi 等文件夹用来存放不同分辨率的图片；layout 文件夹用来存放项目中所用到的 xml 格式的布局文件；menu 文件夹用来存放描述菜单的 xml 文件；values 文件夹用来存放项目所需的字符串等资源。

（6）AndroidManifest.xml 文件为 xml 格式的系统控制文件，也称为项目清单文件，每个 Android 应用程序都有该文件。应用程序所创建的 Activity、Service 以及 Content Provider 等组件都必须在该文件中进行声明或注册。应用程序所需的系统权限也必须在该文件中进行声明。

至此，第一个 Android 应用程序就创建完成了，接下来需要将这个程序运行在一个设备上。设备可以为 Android 手机或 Android 模拟器。这里以模拟器为例运行该程序。由于是第一次运行 Android 项目，所以系统中没有任何模拟器，需要我们自己创建一个。下面详细介绍创建过程。

（1）单击 search 菜单下的按钮，如图 9-15 所示，会弹出一对话框，切换到 Devices Definitions 页面，选择一个合适尺寸的模拟器，如 3.7″WVGA，如图 9-16 所示。

图 9-15　创建模拟器（步骤 1）　　　图 9-16　创建模拟器（步骤 2）

（2）单击右边按钮【create AVD】，弹出一对话框，在 AVD Name 栏输入创建的模拟器名称，如"AVD1"，然后将光标移到 SD Card 栏中 Size 输入框中，输入模拟器中 SD 卡的大小，如 500，单位默认为兆（Mib）。如图 9-17 所示。

（3）单击【为 ok】按钮，模拟器 AVD1 创建完成，在 Android Virtual Devices 选项卡中将会列出该模拟器，如图 9-18 所示。

（4）选中列表中的"AVD1"，单击右边的【start】按钮，弹出一对话框，单击【launch】按钮即可启动模拟器。模拟器启动需要花费一些时间，启动后的效果如图 9-19 所示。

（5）模拟器创建成功并运行后，就可以将之前创建好的 Hello World 程序安装到该模拟器中运行了。选中 Hello World 工程，单击鼠标右键，依次选择 Run As、Android Application，如图 9-20 所示。运行后的效果如图 9-21 所示。

9.2　视频播放器开发

有了前面的基础后，接下来就介绍一个 Android 应用项目的开发过程，以帮助读者进一步了解 Android 项目的开发流程。

视频播放器主要用于个人娱乐或者学习，在连续工作或学习一段时间后播放一个电影或 MV 来放松一下紧张的心情，或者在乘坐公交或地铁时打发一下时间是一个不错的选择。所以，视频播放器的开发有很强的应用价值。

图 9-17　创建模拟器（步骤 3）　　　　图 9-18　创建模拟器（步骤 4）

 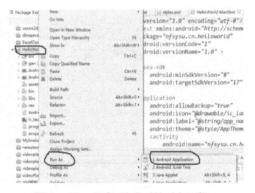

图 9-19　模拟器启动后　　　　图 9-20　运行 Hello World 工程

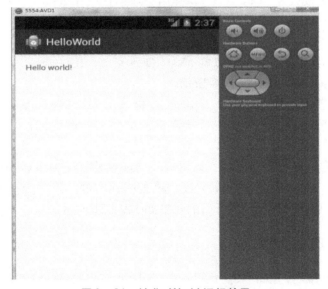

图 9-21　Hello World 运行效果

9.2.1 需求分析

对软件需求的分析是软件开发过程中最重要的步骤之一。一个好的播放器不但要实现播放器的基本功能和扩展功能，还需要有友好的界面。鉴于部分读者可能接触 Android 不久，对 Android 不是很熟悉，所以本章的目的之一是帮助读者在项目开发过程中学会 Android 基本组件和控件，如 TextView、ListView、Button、SimpleCursorAdapter、SQLiteOpenHelper、Activity、Broadcast、LinearLayout、RelativeLayout、VideoView 等的使用，并使用这些基本组件完成一个完整的具有较大参考价值的 Android 应用。视频播放器的流程如图 9-22 所示。结合流程图，接下来对视频播放器从功能需求和界面需求两个方面进行分析。

图 9-22 播放器流程

1. 视频播放器的功能需求

本章开发的视频播放器不但需要实现视频文件的播放及播放控制功能，还需要实现播放列表、异常处理、菜单、播放进度、电量和时间显示等功能。详细介绍如下：

（1）获取待播放的视频并在播放列表中显示所有视频。

（2）支持软件国际化。

（3）支持 mp4、3gp 等多种媒体格式文件的播放。

（4）视频播放列表中应显示视频文件名、视频文件大小，以及播放时长等，并根据不同格式视频文件显示不同的图标。

（5）视频播放时应该全屏且横屏，以增强用户体验。

（6）视频的播放和暂停功能的切换。

（7）视频的快进和快退功能。

（8）视频的上一首和下一首功能。

（9）实现指示播放进度的进度条。

（10）视频播放时可以显示当前已播放的时间及当前播放视频的总时间。

（11）对视频不能播放的情况做出相应处理。

（12）【视频播放控制】按钮及【退出】按钮等控件的显示和隐藏的控制。

（13）可以删除视频文件并同时更新视频列表界面。

（14）支持视频详细信息的查询功能。

（15）视频播放退出功能。

（16）全屏播放时，视频播放界面可以显示当前电量和时间。

2. 视频播放器的界面需求

软件界面是人与计算机之间的媒介，播放器界面的美观程度直接关系到播放器的性能能否充分发挥，关系到使用者能否准确、高效、愉快地使用，所以播放器界面的友好性对于播放器至关重要。

本视频播放器主要界面有两个，第一个是所有视频的列表显示界面，如图 9 - 23 所示。

图 9 - 23 视频列表界面

图 9 - 24 视频播放界面

列表中的每一项都应该有视频文件名、代表视频格式的图片、视频大小以及播放时长等信息。为了直观地显示播放列表中的内容，播放列表背景需设置为白色，字体颜色设置为黑色。长按列表中某个视频可以显示菜单界面。菜单界面中布局有播放、删除及详细信息等菜单项，可以通过单击这几个菜单项对视频进行相应的操作。单击播放列表中某个视频进入播放器主界面，或者显示【视频无法播放】窗口。【视频无法播放】窗口是对视频播放时出现异常情况的处理，设计成对话框的形式，通过设置对话框标题和内容提醒用户无法播放此视频，通过单击对话框中的【确定】按钮播放下一个视频。

第二个主要界面是视频播放界面，如图9-24所示。该界面布局有【播放】窗口、【播放控件】窗口、【退出】按钮窗口、【当前播放视频文件名及电量时间】窗口等。【播放控件】窗口漂浮在播放窗口的下方，窗口的控制栏里布局有切换到上一部视频、快退、播放/暂停、快进、切换到下一部视频，窗口底部布局有进度条和播放时间。【退出】按钮、【当前播放视频文件名及电量时间】窗口布局在播放窗口的最上方。可以单击【退出】按钮退出播放。【播放控件】窗口、【退出】按钮、【当前播放视频文件名及电量时间】窗口在视频开始播放的一段时间后会自动消失，可以通过触摸屏幕来控制这几个窗口的显示和隐藏。

根据上面的分析与描述，总结出视频播放器的需求条款如表9-3所示。

表9-3 视频播放器的需求列表

需求编号	需求描述	解释
Req9-1	获取手机sd卡中的视频并列表显示	自动搜索sd中的视频文件
Req9-2	支持软件国际化	根据手机设置显示相应国家的文字
Req9-3	视频播放列表中应显示视频文件名、视频文件大小，以及播放时长等，并根据不同格式视频文件显示不同的图标	参考图9-23
Req9-4	可以删除视频文件并同时更新视频列表界面	
Req9-5	支持视频详细信息的查询功能	
Req9-6	支持mp4、3gp格式文件的播放	
Req9-7	视频播放时应该横屏，以增强用户体验	
Req9-8	视频的播放和暂停功能的切换	通过一个按钮来控制播放与暂停
Req9-9	视频的上一首和下一首功能	提供相应的按钮来切换
Req9-10	实现指示播放进度的进度条	
Req9-11	视频播放时可以显示当前已播放的时间及当前播放视频的总时间	
Req9-12	对视频不能播放的情况做出相应处理	
Req9-13	视频的快进和快退功能	

续表 9-3

需求编号	需求描述	解释
Req9-14	【视频播放控制】按钮及【退出】按钮等控件的显示和隐藏的控制	
Req9-15	视频播放时可以单击【后退】按钮停止视频播放	
Req9-16	全屏播放时，视频播放界面可以显示当前电量、播放视频的名字以及当前时间	

9.2.2 项目目标

本章的设计目标是开发出一个界面如图 9-23 和图 9-24 的视频播放器，并满足 Req9-1 至 Req9-16 的需求条款。

9.2.3 增量开发计划

经过功能分析，我们已经将项目划分为一个个的小功能，接下来需要制订一个开发计划，按照增量开发的方式实现这个系统。增量开发计划以所要实现的功能和界面为出发点，通过功能的迭代实现过程逐步完成项目所预定的目标。

表 9-4 增量开发计划

增量#	功能实现	实现的需求条款	相关技术（组件）
增量 9-1：搜索视频并列表显示	1. 视频列表实现 2. 可以删除视频文件并同时更新视频列表界面 3. 支持视频详细信息的查询功能 4. 软件国际化	Req9-1 至 req9-5	TextView、ListView、ListView 长按事件处理，Dialog、Activity、SQLite、Popupwindow、LoaderManager、LinearLayout、RelativeLayout、SimpleCursorAdapter
增量 9-2：视频播放及控制功能的实现	1. 支持 mp4、3gp 格式文件的播放 2. 视频播放时全屏且横屏 3. 视频的播放和暂停功能的切换 4. 视频的上一首和下一首功能 5. 实现指示播放进度的进度条 6. 视频播放时可以显示当前已播放的时间及当前播放视频的总时间 7. 对视频不能播放的情况做出相应处理 8. 视频的快进和快退功能 9. 【视频播放控制】按钮及【退出】按钮等控件的显示和隐藏的控制	Req9-6 至 Req9-14	MediaController、LinearLayout、VideoView、SeekBar、RelativeLayout

续表9-4

增量#	功能实现	实现的需求条款	相关技术（组件）
增量9-3：自定义视频播放界面	1. 视频播放画面应提供【后退】按钮，单击【后退】按钮回到视频列表界面 2. 视频播放界面可以显示手机当前电量、播放视频文件的名字以及当前时间	Req9-15至Req9-16	SurfaceView、Handler、Dialog、BroadcastReceiver、View、Thread、按键和触摸事件处理、Java反射技术。

9.3 项目增量开发

本节按照增量开发计划依次实现各个增量，按照增量的实现过程最终实现视频播放器的预定功能。

9.3.1 增量9-1：搜索视频并列表显示

增量9-1实现的功能包括搜索sd卡中的视频文件并实现列表显示，在列表显示界面长按某一项可以弹出相应的选项（比如播放、详细信息、删除等），软件国际化等功能。

参照如图9-23所示的界面布局，可以将增量9-1分为以下几个步骤来实现：

1. 创建Android项目，将图片放入项目文件夹下

首先打开安装好Android SDK的Eclipse软件，然后按照以下步骤创建Android项目。

（1）单击File菜单，选择"New"选项，在弹出的列表中选择"project…"选项，如图9-25所示。

（2）选择"project…"选项后，会弹出一个对话框，选择"Android"文件夹下的"Android Application Project"，如图9-26所示。

图9-25 创建Android工程（步骤1）

图9-26 创建Android工程（步骤2）

（3）接下来弹出一个对话框，要求输入 Application Name、Project Name 和 Package Name，在相应的文本框中输入名称，如图 9-27 所示。

（4）一直单击【Next】按钮，当【Next】按钮变灰色后，单击【Finish】按钮，完成项目的创建。创建完成后，Eclipse 左边项目列表显示区会显示刚刚创建的项目名称，如图 9-28 所示。

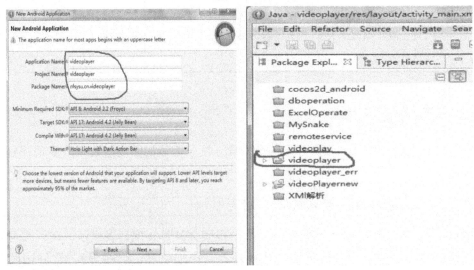

图 9-27　创建 Android 工程（步骤 3）　　　图 9-28　创建 Android 工程（步骤 4）

（5）将项目所需的图片文件（mp4.png 和 threegp.png）拷贝到项目 res 目录中的 drawable-mdpi 子目录下，如图 9-29 所示。

 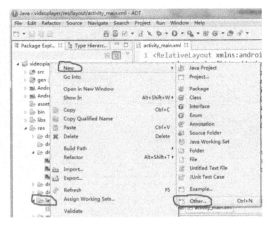

图 9-29　拷贝图片到项目文件夹　　　图 9-30　创建布局文件（步骤 1）

2. 创建用于显示视频列表界面对应的 xml 文件

Android 中一般将界面显示内容放在 xml 中实现，所以为了显示搜索到的视频文件，首先需要创建 xml 文件。用于显示视频列表的 Activity 命名为 VideoBrowserActivity.Java，VideoBrowserActivity 需要一个 xml 布局文件，该布局文件的名字为 video_browser.xml，这个文件目前不存在，所以首先要创建它，步骤如下：

（1）选中项目中 res 目录下的子目录 layout，然后单击鼠标右键，在弹出的列表中依次单击"New"、"Other"，如图 9 – 30 所示。

（2）在弹出的对话框中选择"Android XML Layout File"，如图 9 – 31 所示。

（3）单击【Next】按钮，在弹出的对话框中输入要创建的 XML 文件的名字，如图 9 – 32 所示。

图 9 – 31　创建布局文件（步骤 2）　　图 9 – 32　创建布局文件（步骤 3）

（4）单击【Finish】按钮，res 目录中的子目录 layout 下将生成 video_browser.xml 文件。

布局文件创建好后，接下来需要往 xml 文件中撰写相关代码。由于该布局文件是用来显示视频列表的，所以只需往该文件中添加一个 ListView 控件即可，如代码 9 – 1 所示。

【代码 9 – 1】video_browser.xml 代码

```
<RelativeLayout xmlns:android = "http://schemas.android.
com/apk/res/android"
xmlns:tools = "http://schemas.android.com/tools"
android:layout_width = "match_parent"
android:layout_height = "match_parent"
android:paddingBottom = "@dimen/activity_vertical_margin"
android:paddingLeft = "@dimen/activity_horizontal_margin"
android:paddingRight = "@dimen/activity_horizontal_margin"
android:paddingTop = "@dimen/activity_vertical_margin"
tools:context = ".VideoPlayActivity" >
<ListView
    android:id = "@ +id/listview"
    android:layout_width = "wrap_content"
```

```
        android:layout_height = "wrap_content"
    />
</RelativeLayout>
```

从图 9-23 中可以看出，这个 ListView 中的每一项包含有视频文件类型所对应的图片，视频文件的名字以及播放时长等信息，所以 ListView 中的每一项（item）也需要一个布局文件来描述。该文件命名为：video_item.xml。创建过程跟 video_browser.xml 文件的创建过程一样，这里不再重复。文件创建成功后，需要撰写相关的 XML 代码。该 XML 文件中应包含一个用于显示视频类型的 ImageView 控件，以及三个分别用于显示视频文件的名称、视频文件大小、视频文件播放时长的 TextView 控件。这四个控件需要使用 RelativeLayout、LinearLayout 来包装一下以达到项目所需的界面效果，最终的代码如代码 9-2 所示。

【代码 9-2】 video_item.xml 代码

```xml
<?xml version = "1.0" encoding = "utf-8"? >
<RelativeLayout xmlns:android = "http://schemas.android.com/apk/res/android"
    android:layout_width = "match_parent"
    android:layout_height = "match_parent"
    android:orientation = "horizontal" >
<ImageView
    android:id = "@ +id/imageview"
    android:layout_width = "wrap_content"
    android:layout_height = "wrap_content"
    android:src = "@drawable/ic_launcher" />
<LinearLayout
    android:id = "@ +id/linearlayout"
    android:layout_toRightOf = "@id/imageview"
    android:layout_width = "wrap_content"
    android:layout_height = "wrap_content"
    android:layout_marginLeft = "5dp"
    android:orientation = "vertical" >
<TextView
    android:id = "@ +id/videoname"
    android:layout_width = "wrap_content"
    android:layout_height = "wrap_content"
    android:text = "name" />
<TextView
    android:id = "@ +id/videosize"
    android:layout_width = "wrap_content"
    android:layout_height = "wrap_content"
    android:text = "" />
</LinearLayout>
```

```
<TextView
    android:id = "@ + id/videoduration"
    android:layout_width = "wrap_content"
    android:layout_height = "wrap_content"
    android:layout_alignParentRight = "true"
    android:layout_marginRight = "5dp"
    android:layout_marginTop = "10dp"
    android:text = " " />
</RelativeLayout>
```

3. 创建用于实现软件国际化的资源文件和文件夹

所谓的国际化就是根据手机当前选择的语言（在系统设置中的语言选项中进行设置），显示相应国家的文字。在 Android 中，实现国际化比较简单，只需创建相应的文件夹和文件。在项目的 res 目录下，默认已经创建好了一个 values 文件夹，values 文件夹下有一个 strings.xml 文件。该文件用来存储项目中所用到的字符串资源。一般情况下，Android 系统从 values 文件夹下寻找相关的字符串等资源。如果要显示其他国家和地区的文字，则需创建相应的文件夹和文件。接下来以中国地区简体中文显示为例介绍软件国际化的实现过程。

（1）先选中项目中的 res 目录，然后单击鼠标右键，依次选择"New"、"Other"，弹出一对话框，选择"Android XML Values File"，如图 9-33 所示。

（2）单击【Next】按钮后，在 File 那一栏后面的文本输入框中输入"strings.xml"，如图 9-34 所示。

图 9-33 软件国际化（步骤一）　　　　图 9-34 软件国际化（步骤二）

（3）单击【Next】按钮，选中左边 Available Qualifiers 这一栏的"语 Language"，然后单击【->】按钮，则 Chosen Qualifiers 这一栏中会出现刚刚选择的"语 Language"，并且 Chosen Qualifiers 右边出现新的一栏：Language，单击 Language 栏中的下拉箭头，选择"zh"（zh 代表中文的意思），结果如图 9-35 所示。

（4）仿照步骤三，选中左边 Available Qualifiers 这一栏的"Region"，然后单击【-

>】按钮，在右边新出现的 Region 栏中输入"CN"，然后单击【Finish】按钮。结果如图 9－36 所示。

图 9－35　软件国际化（步骤三）　　　图 9－36　软件国际化（步骤四）

（5）经过以上步骤后，在 res 目录下将生成一个新的文件夹：values－zh－rCN。该文件夹下生成了 strings.xml 文件。这里的"zh"表示中文的意思，"rCN"表示中国地区的意思。结果如图 9－37 所示。

图 9－37　软件国际化（步骤五）

(6) 接下来需要在 values 目录下的 strings.xml 文件中添加项目所需要的字符串资源，添加后的代码如代码 9-3 所示。

(7) 接着需要在 values-zh-rcn 目录下的 strings.xml 文件中添加项目所需要的字符串资源。将步骤六中的 strings.xml 文件中的代码拷贝到 values-zh-rcn 目录下的 strings.xml 文件中，并将字符串的值修改为中文。例如，对于文件中的第一个字符串"app_name"，它在 values 目录下的 strings.xml 文件的值为"Video Player"，在 values-zh-rcn 目录下的 strings.xml 文件中则需将其值改为"视频播放器"，如代码 9-4 所示（注意，只需修改字符串的值，字符串的名字"app_name"不修改）。

【代码 9-3】values 文件夹中的 strings.xml

```xml
<?xml version="1.0" encoding="utf-8"?>
<resources>
    <string name="app_name">Video Player</string>
    <string name="action_settings">Settings</string>
    <string name="total_video_list">total videos</string>
    <string name="play_history_list">play history</string>
    <string name="favorite">favorite</string>
    <stringname="VideoView_error_text_invalid_progressive_playback">invalid progressive playback</string>
    <string name="VideoView_error_text_unknown">Unknown error</string>
    <string name="VideoView_error_button">Error</string>
    <string name="info_selection">Information selection</string>
    <string name="play">play</string>
    <string name="more_info">more information</string>
    <string name="delete">delete</string>
    <string name="confirm_delete">Are you sure to delete?</string>
    <string name="confirm_btn">OK</string>
    <string name="cancle_btn">Cancle</string>
    <string name="delete_succ">Delete file successfully</string>
    <string name="delete_err">Error! Can not delete the file</string>
    <string name="no_video_file_found">no video file found!</string>
</resources>
```

【代码 9-4】values-zh-rcn 文件夹中的 strings.xml

```xml
<?xml version="1.0" encoding="utf-8"?>
<resources>
    <string name="app_name">视频播放器</string>
    <string name="action_settings">设置</string>
    <string name="total_video_list">所有视频</string>
    <string name="play_history_list">播放历史</string>
    <string name="favorite">收藏夹</string>
    <string name="VideoView_error_text_invalid_progressive_playback">无效的进度回放</string>
```

```xml
<string name = "VideoView_error_text_unknown">未知错误</string>
<string name = "VideoView_error_button">错误</string>
<string name = "info_selection">信息选项</string>
<string name = "play">播放</string>
<string name = "more_info">更多信息</string>
<string name = "delete">删除</string>
<string name = "confirm_delete">确定删除吗?</string>
<string name = "confirm_btn">确定</string>
<string name = "cancle_btn">取消</string>
<string name = "delete_succ">已成功删除文件</string>
<string name = "delete_err">删除文件出错</string>
<string name = "no_video_file_found">没有找到视频文件!</string>
</resources>
```

经过以上几步就完成了软件国际化的功能。如果用户在系统设置菜单的语言选项中选择了"English",则该视频播放器相应的文字就会显示为英文的字符串；如果用户在系统设置菜单的语言选项中选择了"简体中文",则该视频播放器相应的文字就会显示为简体中文的文字。效果如图 9-38、图 9-39 所示。目前,该软件只实现了能显示简体中文和英文的文字,如果需要显示其他国家的语言文字,比如西班牙语、俄语等,则需要仿照上述步骤创建相应的文件夹和修改 strings.xml 文件中字符串的值。

图 9-38　系统语言为英语时的显示效果　　图 9-39　系统语言为中文时的显示效果

4. 搜索 sd 卡中的视频文件并列表显示

现在开始需要编写相关代码搜索手机 sd 卡中的视频文件，并将这些文件的相关信息显示在 listview 当中。视频文件有一些基本信息，如文件名、文件路径、文件类型、文件大小、播放时长等，为方便管理和显示视频文件，需要创建一个 JavaBean 来描述视频文件。该 JavaBean 的名字为 VideoInfo，放在 nfsysu. cn. videoplayer. domain 包下，它有 6 个属性，每个属性都有相应的 get 和 set 方法，完整代码如代码 9-5 所示。

程序需要将搜索到的视频文件存放到应用程序自己的数据库文件中，所以需要建立相关的类来建立数据库和对数据库进行增、删、改、查。首先需要创建数据库"video. db"，它有一个表"video"。根据 VideoInfo 的 JavaBean 属性，可以得到 video 表具有的几个字段：id、name、size、type、duration、path。如表 9-5 所示。注意，由于 size 字段含有"KB"、"MB"等字符，为了方便起见，也将其类别设置为 varchar 而不是 int 类型。

表 9-5 视频信息 video

字段名称	字段名称	类别	主键	非空
视频 ID	id	int	是	是
视频名称	name	varchar	否	是
视频大小	size	varchar	否	是
视频类型	type	varchar	否	是
视频时长	duration	varchar	否	是
视频路径	path	varchar	否	是

【代码 9-5】VideoInfo 代码

```
package nfsysu.cn.videoplayer.domain;
public class VideoInfo {
    private Integer id; //id
    private String name; //名称
    private String size; //大小
    private String type; //类型
    private String duration; //时长
    private String path; //路径
    public VideoInfo(String name, String size, String type, String duration, String path) {
        this.name = name;
        this.size = size;
        this.type = type;
        this.duration = duration;
        this.path = path;
    }
```

```java
        public VideoInfo() {
        }
        public VideoInfo(Integer id, String name, String size, String type,String duration, String path) {
            super();
            this.id = id;
            this.name = name;
            this.size = size;
            this.type = type;
            this.duration = duration;
            this.path = path;
        }
        public Integer getId() {
            return id;
        }
        public void setId(Integer id) {
            this.id = id;
        }
        public String getName() {
            return name;
        }
        public void setName(String name) {
            this.name = name;
        }
        public String getSize() {
            return size;
        }
        public void setSize(String size) {
            this.size = size;
        }
    public String getType() {
            return type;
        }
        public void setType(String type) {
            this.type = type;
        }
        public String getDuration() {
            return duration;
        }
        public void setDuration(String duration) {
            this.duration = duration;
        }
```

```java
    public String getPath() {
        return path;
    }
    public void setPath(String path) {
        this.path = path;
    }
}
```

接下来需要创建类 DBOpenHelper，该类继承了 SQLiteOpenHelper，用来创建视频文件的数据库。该类放在 nfsysu.cn.videoplayer.service 包下面，如代码 9-6 所示。

【代码 9-6】类 DBOpenHelper 代码

```java
package nfsysu.cn.videoplayer.service;
import android.content.Context;
import android.database.sqlite.SQLiteDatabase;
import android.database.sqlite.SQLiteOpenHelper;
public class DBOpenHelper extends SQLiteOpenHelper {
    private String name = "video.db";
    private int version = 1;
    public DBOpenHelper(Context context) {
        super(context, "video.db", null, 1);
    }
    @Override
    public void onCreate(SQLiteDatabase db) { //数据库第一次被创建时调用
        String sql = "create table video(id integer primary key, name varchar(255) not null," + "size varchar(100), type varchar(10), duration varchar(100), path varchar(255) not null)";
        db.execSQL(sql);
    }
    @Override
    public void onUpgrade(SQLiteDatabase db, int oldVersion, int newVersion) {
    }
}
```

同时，也需要创建一个服务类来对视频文件进行增删改查等操作，该服务类的名字为 VideoDBService，同样放在 nfsysu.cn.videoplayer.service 包下，该类的代码如代码 9-7 所示。

【代码 9-7】类 VideoDBService 代码

```java
package nfsysu.cn.videoplayer.service;
import Java.util.ArrayList;
import Java.util.List;
import nfsysu.cn.videoplayer.domain.VideoInfo;
import android.content.Context;
import android.database.Cursor;
```

```java
import android.database.sqlite.SQLiteDatabase;
public class VideoDBService {
    private DBOpenHelper dbhelper;
    public VideoDBService(Context context) {
        this.dbhelper = new DBOpenHelper(context);
    }
    public void save(VideoInfo video){
        SQLiteDatabase db = dbhelper.getWritableDatabase();
        db.execSQL("insert into video(id,name,size,type,duration,path) values(?,?,?,?,?,?)", new Object[]{video.getId(), video.getName(), video.getSize(), video.getType(), video.getDuration(), video.getPath()});
    }
    public void delete(Integer id){
        SQLiteDatabase db = dbhelper.getWritableDatabase();
        db.execSQL("delete from video where id = ?", new Object[]{id});
    }
    public void deleteAll( ){
        SQLiteDatabase db = dbhelper.getWritableDatabase();
        db.execSQL("delete from video");
    }
    public void update(VideoInfo video){
        SQLiteDatabase db = dbhelper.getWritableDatabase();
        db.execSQL("update video set name = ?, size = ?, type = ?, duration = ?, path = ? where id = ?", new Object[]{video.getName(), video.getSize(), video.getType(), video.getDuration(), video.getPath(), video.getId()});
    }
    public List<VideoInfo> getScrollData(int offset, int max){
        SQLiteDatabase db = dbhelper.getReadableDatabase();
        List<VideoInfo> list = new ArrayList<VideoInfo>();
        VideoInfo videoDesc = null;
        Cursor cursor = db.rawQuery("select * from video order by id asc limit ?,?", new String[]{String.valueOf(offset), String.valueOf(max)});
        while(cursor.moveToNext()){
            int videoid = cursor.getInt(cursor.getColumnIndex("id"));
            String size = cursor.getString(cursor.getColumnIndex("size"));
            String duration = cursor.getString(cursor.getColumnIndex("duration"));
            String name = cursor.getString(cursor.getColumnIndex("name"));
            String type = cursor.getString(cursor.getColumnIndex("type"));
            String path = cursor.getString(cursor.getColumnIndex("path"));
            videoDesc = new VideoInfo(videoid, name, size, type, duration, path);
```

```java
            list.add(videoDesc);
    }
        cursor.close();
    return list;
}
    public Cursor getCursorScrollData(int offset, int max){
        SQLiteDatabase db = dbhelper.getReadableDatabase();
    Cursor cursor = db.rawQuery("select id as _id, name,size,path,duration,type from video order by id asc limit ?,?", new String[]{String.valueOf(offset), String.valueOf(max)});
        return cursor;
}
    public int getCount(){
        SQLiteDatabase db = dbhelper.getReadableDatabase();
        Cursor cursor = db.rawQuery("select count(*) from video", null);
        cursor.moveToFirst();
        int count = cursor.getInt(0);
        cursor.close();
        return count;
}
}
```

由于 ListView 显示数据（本例为视频文件）信息需要一个适配器（Adapter），所以创建一个适配器类 VideoAdapter，该类继承 SimpleCursorAdapter 类，放在 nfsysu.cn.videoplayer.adapter 包下。由于用到了 SimpleCursorAdapter 类，需要修改项目 AndroidManifest.xml 文件中最低的 sdk 级别为 11，类 VideoAdapter 代码如代码 9 – 8 所示。

【代码 9 – 8】类 VideoAdapter 代码

```java
package nfsysu.cn.videoplayer.adapter;
import nfsysu.cn.videoplayer.R;
import android.content.Context;
import android.database.Cursor;
import android.view.View;
import android.widget.ImageView;
import android.widget.SimpleCursorAdapter;
public class VideoAdapter extends SimpleCursorAdapter {
    private Context mContext;
    public VideoAdapter(Context context, int layout, Cursor c, String[] from, int[] to, int flags) {
        super(context, layout, c, from, to, flags);
        mContext = context;
    }
```

```java
    @Override
    public void bindView(View view, Context context, Cursor cursor) {
        super.bindView(view, context, cursor);
        ImageView imageView = (ImageView) view.findViewById(R.id.imageview);
        String type = cursor.getString(cursor.getColumnIndex("type"));
        if("mp4".equalsIgnoreCase(type)){
            imageView.setImageResource(R.drawable.mp4);
        }else if("3gp".equalsIgnoreCase(type)){
            imageView.setImageResource(R.drawable.threegp);
        }
    }
    @Override
    public void setViewImage(ImageView v, String value) {
        super.setViewImage(v, value);
    }
}
}
```

接下来，找到项目源文件中的 MainActivity.Java 文件（创建项目时默认的文件名），将其重命名（选中文件后单击右键，依次选择"Refactor"、"Rename"）为 VideoBrowserActivity.Java，该类继承 Activity 类，用来显示视频列表。该类的部分代码如代码 9-9 所示。

【代码 9-9】类 VideoBrowserActivity 部分代码

```java
package nfsysu.cn.videoplayer;
import Java.io.File; //其他 import 语句省略,读者可以按快捷键 Alt + Shift + O 导入相关包
public class VideoBrowserActivity extends Activity{
    private static final String TAG = "VideoBrowserActivity";
    private ListView listView;
    private VideoAdapter videoAdapter;
    private static VideoBrowserActivity instance;
    private VideoDBService vds;
    private int VideoCount;
    private StringBuilder mFormatBuilder;
    private Formatter mFormatter;
    private static final int PLAY = 0;
    private static final int MORE_INFO = 1;
    private static final int DELETE = 2;
```

该类中有如下一些方法和内部类。

onCreate() 方法。该方法在 VideoBrowserActivity 类的实例对象被创建时调用，在该方法中主要完成一些初始化的工作。如代码 9-10 所示。

【代码 9 - 10】类 VideoBrowserActivity 中的 onCreate () 方法

```
@Override
protected void onCreate(Bundle savedInstanceState){
    super.onCreate(savedInstanceState);
    instance = this;
    vds = new VideoDBService(this);
    setContentView(R.layout.video_browser);
    mFormatBuilder = new StringBuilder();
    mFormatter = new Formatter(mFormatBuilder, Locale.getDefault());
    listView = (ListView) this.findViewById(R.id.listview);
    this.getLoaderManager().initLoader(0, null, new MyLoaderCallback());
    bindData();
    listView.setOnItemClickListener(new ItemClickListener());
    listView.setOnCreateContextMenuListener(new MyOnCreateContextMenuListener());
}
```

在 onCreate () 方法中调用了 getLoaderManager () 方法获取 LoaderManager 对象, 然后调用该对象的 initLoader () 方法, 用来加载视频文件。initLoader () 方法的第一个参数为 Loder 的 ID（唯一标识）, 一般设为 0; 第二个参数是给 Loader 初始化的时候传递的参数［也就是 onCreateLoade () 中的第二个参数］。第三个参数为一个需要实现 LoaderManager.LoaderCallbacks < Cursor > 接口的对象。所以程序创建一个内部类 MyLoaderCallback, 该类实现了 LoaderManager.LoaderCallbacks < Cursor > 接口。该内部类的代码如代码 9 - 11 所示。

【代码 9 - 11】内部类 MyLoaderCallback

```
private class MyLoaderCallback implements LoaderManager.LoaderCallbacks<Cursor>{
    public Loader<Cursor> onCreateLoader(int id, Bundle args){
        CursorLoader cursor = new CursorLoader(VideoBrowserActivity.this,
        MediaStore.Video.Media.EXTERNAL_CONTENT_URI, null, null, null, null);
            Log.i(TAG, "onCreateLoader()");
            return cursor;
    }
    public void onLoadFinished(Loader<Cursor> loader, Cursor data){
        VideoCount = data.getCount();
        Log.i(TAG, "VideoCount: " + VideoCount);
        if(!data.moveToFirst()){
    Toast.makeText(VideoBrowserActivity.this, getString(R.string.no_video_file_found), Toast.LENGTH_LONG).show();
        }
        else{
            vds.deleteAll();
```

```java
            do {
                VideoInfo vi = new VideoInfo();
                    vi.setId ( data.getInt ( data.getColumnIndex
(MediaStore.Video.Media._ID)));
            vi.setName ( data.getString ( data.getColumnIndex
(MediaStore.Video.Media.DISPLAY_NAME)));
            vi.setPath ( data.getString ( data.getColumnIndex
(MediaStore.Video.Media.DATA)));
                int size = data.getInt ( data.getColumnIndex
(MediaStore.Video.Media.SIZE));
        String sizeStr = ByteConvert2GBMBKB(size);
        vi.setSize(sizeStr);
                int duration = data.getInt ( data.getColumnIndex
(MediaStore.Video.Media.DURATION));
        String durationStr = millSectoHourMinSec(duration);
        vi.setDuration(durationStr);
            String path = data.getString ( data.getColumnIndex
(MediaStore.Video.Media.DATA));
        String type = path.substring(path.lastIndexOf(".") + 1);
        vi.setType(type);
        vds.save(vi);
        } while (data.moveToNext());
    }
    Log.i(TAG, "vds count:" + vds.getCount());
    }
    public void onLoaderReset(Loader < Cursor > loader) {
            videoAdapter.swapCursor(null);
    }
  }
```

在内部类 MyLoaderCallback 的 onCreateLoader () 方法中创建 CursorLoader 对象并返回。CursorLoader 对象创建后就可以加载（load）数据了。当加载数据结束的时候（第一次读取数据或者数据有改变的时候会加载数据）会调用 onLoadFinished () 方法，而 onLoaderReset () 方法只有在销毁一个 loader 的时候才会被调用。在 onLoadFinished () 方法中，将查询到的视频文件通过 VideoDBService 类提供的 save () 方法保存到数据库中。

（1）接下来，在 onCreate () 方法中调用了 bindData () 方法，该方法的主要作用是将数据库中的数据通过适配器（VideoAdapter）显示在 ListView 中。如代码 9 – 12 所示。

【代码 9 – 12】 bindData () 方法
```java
    public void bindData() {
        String[] from = new String[] { "name", "size", "duration" };
```

```
    int[] to = { R.id.videoname, R.id.videosize, R.id.videoduration };
    Cursor cursor = vds.getCursorScrollData(0, vds.getCount());
    cursor.moveToFirst();
    videoAdapter = new VideoAdapter(this,
            R.layout.video_item, cursor, from, to, 0);
    Log.i(TAG, "curor = " + cursor);
    listView.setAdapter(videoAdapter);
}
```

(2) 在 onCreate() 方法中通过 listView.setOnItemClickListener() 方法给 ListView 设置了 ListView 条目 (item) 被单击时的监听器。该方法需要一个实现了 OnItemClickListener 接口的对象, 所以程序创建了一个内部类 ItemClickListener, 该类实现了 OnItemClickListener 接口。如代码 9 – 13 所示。

【代码 9 – 13】内部类 ItemClickListener

```
private class ItemClickListener implements OnItemClickListener {
    public void onItemClick(AdapterView <? > parent, View view, int position, long id) {
        Cursor cursor = vds.getCursorScrollData(0, vds.getCount());
        if (cursor ! = null) {
            cursor.moveToPosition(position);
            String path = cursor.getString(cursor.getColumnIndex("path"));
            Log.i(TAG, "path: " + path);
            Intent intent = new Intent (VideoBrowserActivity.this, VideoPlayActivity.class);
            intent.putExtra("path", path);
            intent.putExtra("position", position);
            startActivity(intent);
        }
    }
}
```

当 ListView 中的某个条目被单击后, 该内部类的 onItemClick() 方法将被调用。在该方法内部, 将单击时的位置 (position) 和视频文件的路径 (path) 通过 intent 对象传递给 VideoPlayActivity, VideoPlayActivity 收到该 intent 进行解析后就可以播放相应的视频文件了。

(3) 在 onCreate() 方法中通过 listView.setOnCreateContextMenuListener() 方法给 ListView 设置了条目 (item) 被长按时的监听器对象。该方法需要一个实现 OnCreateContextMenuListener 接口的对象参数, 程序创建了一个内部类 MyOnCreateContextMenuListener, 该类实现了 OnCreateContextMenuListener 接口。如代码 9 – 14 所示。其中, PLAY、MORE_INFO 和 DELETE 是刚开始时定义的常量, 值分别为 0、1、2。

【代码 9 – 14】内部类 MyOnCreateContextMenuListener

```
private final class MyOnCreateContextMenuListener implements
```

```
OnCreateContextMenuListener{
  public void onCreateContextMenu(ContextMenu menu,
      View v, ContextMenuInfo menuInfo) {
    menu.setHeaderTitle(R.string.info_selection);
    menu.add(0, PLAY, 0, R.string.play);
    menu.add(0, MORE_INFO, 0, R.string.more_info);
    menu.add(0, DELETE, 0, R.string.delete);
  }
}
```

当长按 ListView 的某一项时,将会弹出如图 9-40 所示的菜单。

图 9-40　长按菜单　　　　图 9-41　单击删除选项后

用户可以选择"播放"、"更多信息"或者"删除"。当用户选择这些选项时,onContextItemSelected()方法将被调用,所以需要重写该方法,如代码 9-15 所示。

【代码 9-15】重写 onContextItemSelected()方法

```
@Override
public boolean onContextItemSelected(MenuItem item) {
  AdapterContextMenuInfo menuInfo = (AdapterContextMenuInfo) item.getMenuInfo();
    final int position = menuInfo.position;
    final Cursor cursor = vds.getCursorScrollData(0, vds.getCount());
    switch(item.getItemId()) {
    case PLAY:
      if(cursor != null) {
        cursor.moveToPosition(position);
        String path = cursor.getString(cursor.getColumnIndex("path"));
```

```java
            Log.i(TAG, "path: " + path);
            Intent intent = new Intent(VideoBrowserActivity.this,
                VideoPlayActivity.class);
            intent.putExtra("path", path);
            intent.putExtra("position", position);
            startActivity(intent);
            }
            case MORE_INFO:
              if(cursor ! = null){
                showMoreInfo(cursor, position);
              }
              break;
        case DELETE:
            new AlertDialog.Builder(this).setTitle(R.string.delete)
            .setMessage(getString(R.string.confirm_delete))
                .setPositiveButton ( R.string.confirm _ btn, new DialogInterface.OnClickListener(){
                    public void onClick(DialogInterface dialog, int which) {
                    if(cursor ! = null){
                    cursor.moveToPosition(position);
                    String path = cursor.getString(cursor.getColumnIndex("path"));
                    Log.i(TAG, "path: " + path);
                    int id = cursor.getInt(cursor.getColumnIndex("_id"));
                    vds.delete(id);
                    File file = new File(path);
                    boolean result = file.delete();
                    if(result){
                        videoAdapter.notifyDataSetChanged();
                            videoAdapter.swapCursor ( vds.getCursorScrollData ( 0,
vds.getCount()));
                        listView.setAdapter(videoAdapter);
            Toast.makeText(VideoBrowserActivity.this, getString(R.string.delete_
succ) + path, Toast.LENGTH_LONG).show();
                    }
                    else{
                            Toast.makeText ( VideoBrowserActivity.this, getString
(R.string.delete_err) + path, Toast.LENGTH_LONG).show();
                    }
                    }
                    }
                }
            })
            .setNegativeButton(R.string.cancle_btn, new DialogInterface. OnClickLis-
```

```
tener() {
        public void onClick(DialogInterface dialog, int which) {
            dialog.dismiss();
        }
    }).show();
    break;
    default:
    break;
    }
    return super.onContextItemSelected(item);
}
```

当用户单击【播放】时,"case PLAY"这一部分代码被执行。该部分代码将选择的视频文件的路径和位置通过 intent 传递给 VideoPlayActivity,VideoPlayActivity 解析后即可进行播放。

当用户单击【删除】时,"case DELETE"这一部分代码将被执行,弹出一对话框,提示用户是否确定删除该视频文件,如图 9-41 所示。该对话框有两个按钮,分别是【取消】和【确定】,在代码部分对这两个按钮进行监听。如代码 9-15 所示。

当用户单击【更多信息】选项时,"case MORE_ INFO"这一部分代码被执行,显示所选择的视频文件的名称、大小、路径、时长等信息。如图 9-42 所示。这时,showMoreInfo()方法将被调用。showMoreInfo()方法中将创建一个 PopupWindow 用来显示视频信息,如代码 9-16 所示。

图 9-42 单击"更多信息"后效果

【代码 9 – 16】 showMoreInfo（ ）方法

```java
private void showMoreInfo(Cursor cursor, int position) {
  if(cursor != null){
    cursor.moveToPosition(position);
    String name = cursor.getString(cursor.getColumnIndex("name"));
    String size = cursor.getString(cursor.getColumnIndex("size"));
    String path = cursor.getString(cursor.getColumnIndex("path"));
    String duration = cursor.getString(cursor.getColumnIndex("duration"));
    String type = cursor.getString(cursor.getColumnIndex("type"));
    StringBuffer sb = new StringBuffer();
    sb.append("名称:" + name + "\r\n");
    sb.append("大小:" + size + "\r\n");
    sb.append("路径:" + path + "\r\n");
    sb.append("播放时长:" + duration + "\r\n");
    sb.append("类型:" + type + "\r\n");
    View popView;
    LayoutInflater layoutInflater = getLayoutInflater();
    popView = layoutInflater.inflate(R.layout.moreinfo_popupwindow, null);
    final PopupWindow popwin = new PopupWindow(popView,LayoutParams.MATCH_PARENT,LayoutParams.WRAP_CONTENT,true);
    TextView contentTv = (TextView)popView.findViewById(R.id.tv);
    contentTv.setText(sb.toString());
    popwin.setOutsideTouchable(true);
    Button confirmBtn = (Button) popView.findViewById(R.id.confirm_btn);
    confirmBtn.setOnClickListener(new OnClickListener() {
      public void onClick(View v) {
        if(popwin.isShowing()){
          popwin.dismiss();
        }
      }
    });
    ColorDrawable dw = new ColorDrawable(0000000000);
    popwin.setBackgroundDrawable(dw);
    popwin.showAtLocation(this.findViewById(R.id.listview), Gravity.CENTER,0,0);
  }
}
```

通过 LoaderManager 对象加载的视频文件的相关信息如文件大小、播放时长的单位分别是 B（byte 字节）和 ms（毫秒），这两个单位太小，直接显示的话用户体验不大好，所以我们需要提供相应的方法将其转换为合适的单位。程序提供了 ByteConvert2GBMBKB（int size）方法将 B（字节）转换为 kB（千字节）、MB（兆字节）或者

GB（吉字节）；millSectoHourMinSec（int duration）方法用于将 ms（毫秒）转换为小时、分钟和秒等单位。如代码 9 – 17 和代码 9 – 18 所示。

【代码 9 – 17】 ByteConvert2GBMBKB（） 方法

```java
public String ByteConvert2GBMBKB(int size) {
    int b = size % 1024;
    int kb = (size /1024) % 1024;
    int mb = size /(1024 * 1024) % 1024;
    int gb = size /(1024 * 1024 * 1024);
    String sizeStr;
    mFormatBuilder.setLength(0);
    if(gb > 0) {
        sizeStr = mFormatter.format("% 02d.% d", gb, mb /102) + "GB";
    } else if(mb > 0) {
        sizeStr = mFormatter.format("% 02d.% d", mb, kb /102).toString() + "MB";
    } else if(kb > 0) {
        sizeStr = mFormatter.format("% 02d.% d", kb, b /102).toString() + "KB";
    } else {
        sizeStr = size + "B";
    }
    return sizeStr;
}
```

【代码 9 – 18】 millSectoHourMinSec（） 方法

```java
public String millSectoHourMinSec(int duration) {
    int totalSeconds = duration /1000;
    int seconds = totalSeconds % 60;
    int minutes = (totalSeconds /60) % 60;
    int hours = totalSeconds /3600;
    String durationStr;
    mFormatBuilder.setLength(0);
    if(hours > 0) {
        durationStr = mFormatter.format("% d:% 02d:% 02d", hours, minutes, seconds).toString();
    }else{
        durationStr = mFormatter.format("% 02d:% 02d", minutes, seconds).toString();
    }
    return durationStr;
}
```

9.3.2 增量 9-2：视频播放及控制功能的实现

通过增量 9-1，我们已经将视频文件从 sd 卡中搜索出来并显示在列表中，增量 9-2 将在 9-1 的基础上实现视频播放和相应的控制功能。

视频播放界面也需要一个 Activity，所以我们首先创建 VideoPlayActivity.Java 文件，该类同样继承 Activity 类，放在 nfsysu.cn.videoplayer 包下。这里不要忘记的是，创建完 VideoPlayActivity 后，需要在项目清单文件 AndroidManifest.xml 中注册这个 Activity，否则运行时将报错，提示找不到类 VideoPlayActivity。

视频播放界面同样需要一个布局文件来布局视频播放画面，该文件命名为 videoplayer.xml，放在项目的 layout 目录下。Android 为了方便开发者开发视频相关的功能，提供了类 VideoView 和 MediaController，我们可以很方便地利用这两个类进行视频播放和控制。视频播放界面我们只需要一个 VideoView 控件即可，如代码 9-19 所示。

【代码 9-19】 videoplayer.xml 代码

```xml
<?xml version = "1.0" encoding = "utf-8"?>
<LinearLayout xmlns:android = "http://schemas.android.com/apk/res/android"
    android:layout_width = "match_parent"
    android:layout_height = "match_parent"
    android:baselineAligned = "false"
    android:background = "#000000" >
    <VideoView
        android:id = "@ +id/videoview"
        android:layout_width = "match_parent"
        android:layout_height = "match_parent"
        android:layout_gravity = "center_horizontal"
        />
</LinearLayout>
```

视频播放的布局文件已经创建完毕，接下来需编写 VideoPlayActivity.Java 文件，这个类的代码比较简单，我们只需实现 oncreate() 方法即可。如代码 9-20 所示。

【代码 9-20】 VideoPlayActivity.Java 代码

```java
package nfsysu.cn.videoplayer;
import nf.sysu.videoplayer.service.VideoDBService;
import android.app.Activity;
import android.database.Cursor;
import android.net.Uri;
import android.os.Bundle;
import android.util.Log;
import android.view.Menu;
import android.view.View;
import android.view.View.OnClickListener;
import android.view.Window;
```

```java
import android.view.WindowManager;
import android.widget.MediaController;
import android.widget.VideoView;
public class VideoPlayActivity extends Activity {
    private VideoView videoView;
    private VideoDBService vds;
    private int position;
    @Override
    protected void onCreate(Bundle savedInstanceState) {
        super.onCreate(savedInstanceState);
        requestWindowFeature(Window.FEATURE_NO_TITLE); getWindow().setFlags(WindowManager.LayoutParams.FLAG_FULLSCREEN, WindowManager.LayoutParams.FLAG_FULLSCREEN);
        setContentView(R.layout.videoplayer);
        vds = new VideoDBService(this);
        MediaController mediaController = new MediaController(this);
        videoView = (VideoView) this.findViewById(R.id.videoview);
        mediaController.setAnchorView(videoView);
        mediaController.setPrevNextListeners(new OnClickListener() {
            public void onClick(View v) {
                Cursor cursor = vds.getCursorScrollData(0, vds.getCount());
                if(position == vds.getCount() -1){
                    position = 0;
                }else{
                    position++;
                }
                cursor.moveToPosition(position);
                String path = cursor.getString(cursor.getColumnIndex("path"));
                videoView.setVideoURI(Uri.parse(path));
                videoView.start();
            }
        }, new View.OnClickListener() {
            public void onClick(View v) {
                Cursor cursor = vds.getCursorScrollData(0, vds.getCount());
                if(position == 0){
                    position = vds.getCount() -1;
                }else{
                    position--;
                }
                cursor.moveToPosition(position);
                String path = cursor.getString(cursor.getColumnIndex
```

```
("path"));
                    videoView.setVideoURI(Uri.parse(path));
                    videoView.start();
                }
            });
            videoView.setMediaController(mediaController);
            String path = this.getIntent().getStringExtra("path");
            position = this.getIntent().getIntExtra("position",0);
            videoView.setVideoURI(Uri.parse(path));
            videoView.start();
        }
    }
```

视频播放时需要全屏,并且是横屏,onCreate()方法的第2~4行代码即是将播放界面设置为全屏。要设置为横屏,需要在项目清单文件AndroidManifest.xml中注册VideoPlayActivity这个Activity的相应地方加上一句代码:android:screenOrientation ="landscape"即可实现横屏的效果。在onCreate()方法中我们创建了一个MediaController对象mediaController,该对象用于对视频进行播放控制,如快进快退、播放暂停、下一首、上一首等功能。默认情况下【上一首】、【下一首】这两个控制按钮并不会出现在播放界面上,这是因为没有给【下一首】、【上一首】按钮设置监听器对象。所以为了显示这两个按钮,我们通过mediaController对象的setPrevNextListeners()方法给其设置了监听器对象,如代码9-20所示。onCreate()方法的最后4行代码从intent对象中获取要播放的视频的路径,然后调用videoView对象的setVideoURI()方法和start()方法即可播放相应的视频。效果如图9-43所示。

图9-43 视频播放界面

从图9-43可以看出,播放界面有5个按钮,从左到右分别用来实现【上一首】、

【快退】、【播放/暂停】、【快进】、【下一首】功能，在播放界面的最下面还有一个进度条，用来指示当前的播放进度。进度条两边各有一个 TextView，分别用来显示视频已播放的时间和视频总时间。当不触摸屏幕超过三秒，这5个按钮及进度条等均会自动隐藏。由此可见，通过 Android 自带的 VideoView 和 MediaController，我们很容易地就实现了增量 9-2 所要实现的功能。

9.3.3 增量 9-3 自定义视频播放界面

实现增量 9-1 和增量 9-2 后，我们已经可以正常播放 3GP 或 MP4 格式的视频文件了，并且在播放时可以对视频进行播放控制。但是从图 9-43 可以看出，视频播放界面的布局以及播放控制按钮都是 Android 系统默认的，不能满足个性化的需求。为实现更富个性化和更好用户体验的播放界面，我们需要在原来的基础上进行修改。这就需要我们重写 VideoView 和 MediaController 这两个类。我们将新的类命名为 MyVideoView 和 MyMediaController。在项目的 src 目录下分别创建 MyVideoView 和 MyMediaController 两个类，放在 nfsysu.cn.videoplayer.view 包下。我们将在 VideoView 和 MediaController 这两个类的基础上编写 MyVideoView 和 MyMediaController 两个类的代码，这样将是最简单和高效的。将 Android 系统自带的 VideoView.Java 文件的代码拷贝到 MyVideoView.Java 文件中。系统自带的 VideoView 类要实现 MediaPlayerControl、SubtitleController.Anchor 两个接口，但是 SubtitleController.Anchor 这个接口我们不能使用，所以为了编译通过，我们只让 MyVideoView 实现 MediaPlayerControl 这一个接口，然后将与接口 SubtitleController.Anchor 相关的代码都注释掉。另一个需要修改的地方是 VideoView 类中创建 mPreparedListener 对象时用到了 Metadata 这个类，这个类同样不对开发者开放，所以我们也用不了。为了编译通过，我们将与 Metadata 相关的代码都注释掉，直接给变量 mCanPause、mCanSeekBack、mCanSeekForward 赋值为 true，如代码 9-21 所示。

【代码 9-21】MyVideoView.Java 部分代码

```
public class MyVideoView extends SurfaceView implements MediaPlayerControl {
    …//这里省略与 VideoView 类相同的代码
    MediaPlayer.OnPreparedListener mPreparedListener = new MediaPlayer.OnPreparedListener() {
        public void onPrepared(MediaPlayer mp) {
            mCurrentState = STATE_PREPARED;
            //Get the capabilities of the player for this stream
            //由于编译问题,暂时屏蔽,直接给三个变量赋值为 true;
            mCanPause = mCanSeekBack = mCanSeekForward = true;
            /* Metadata data = mp.getMetadata(MediaPlayer.METADATA_ALL,
                    MediaPlayer.BYPASS_METADATA_FILTER);
            if (data ! = null) {
                mCanPause = ! data.has(Metadata.PAUSE_AVAILABLE)
                    ||data.getBoolean(Metadata.PAUSE_AVAILABLE);
```

```
    mCanSeekBack = ! data.has(Metadata.SEEK_BACKWARD_AVAILABLE)
       || data.getBoolean(Metadata.SEEK_BACKWARD_AVAILABLE);
    mCanSeekForward = ! data.has(Metadata.SEEK_FORWARD_AVAILABLE)
       || data.getBoolean(Metadata.SEEK_FORWARD_AVAILABLE);
   } else {
       mCanPause = mCanSeekBack = mCanSeekForward = true;
    } */
       …//后续代码与 VideoView 中的一致,这里省略
    };
}
```

经过这样处理后,MyVideoView 就算改造好了。这里需要注意一点的是,由于改造过程中用到了 API 14 的方法,所以需要在项目清单文件 AndroidManifest.xml 中将最低的 SDK 版本由 11 改为 14 (即 android:minSdkVersion = " 14")。

图 9 – 44　播放控制按钮

MyMediaController 类的改造相对比较复杂,在改造之前,我们需要将自定义的播放控制按钮图片拷贝到项目中,即将图 9 – 44 的六个按钮拷贝到项目的 drawable – mdpi 文件夹下。由于增量 9 – 3 需要在视频播放时显示手机当前电量和提供后退按钮,所以也需要把这些图片拷贝到项目 drawable – mdpi 文件夹中。这里我们将手机电量分为 5 个级别,分别用一张图片来表示,如图 9 – 45 所示。

图 9 – 45　后退按钮和电池电量图片

图片拷贝完毕,接下来需要布局视频播放界面,参看图 9 – 24 可以看出,视频播放界面最上面从左到右是由一个 ImageButton (后退按钮)、2 个 TextView (用来显示当前播放的视频文件名字和显示系统时间) 和一个 ImageView (显示电池电量对应的图片) 组成。所以可以将这四个控件用一个相对布局 (RelativeLayout) 组合在一起。

接下来最左边是一个显示当前已播放时间的 TextView,中间是一个进度条 (Seek-Bar),最右边是一个显示视频总播放时间的 TextView。这三个控件可以用一个水平方向的线性布局 (LinearLayout) 来组合。播放界面的最下面是 5 个播放控制按钮 (Image-Button),它们也用一个水平方向的线性布局来组合。根据上面的描述,我们就可以写出播放界面的布局代码了,我们将这些代码放在 layout 文件夹下的 media.xml 文件中,如代码 9 – 22 所示。

【代码 9 – 22】media.xml 代码

```
<? xml version = "1.0" encoding = "utf-8"? >
<RelativeLayout xmlns:android = "http://schemas.android.com
```

```xml
/apk/res/android"
    android:layout_width = "match_parent"
    android:layout_height = "match_parent"
    android:orientation = "vertical" >
<RelativeLayout
    android:id = "@+id/system_status"
    android:layout_width = "match_parent"
    android:layout_height = "wrap_content"
    android:background = "@drawable/media_bg"
    android:layout_alignParentTop = "true" >
<ImageButton
    android:id = "@+id/back"
    android:layout_width = "wrap_content"
    android:layout_height = "wrap_content"
    android:layout_marginLeft = "10dp"
    android:layout_marginTop = "5dp"
    android:src = "@drawable/back" />
<TextView
    android:id = "@+id/playingvideoname"
    android:layout_width = "wrap_content"
    android:layout_height = "wrap_content"
    android:layout_marginLeft = "10dp"
    android:layout_marginTop = "10dp"
    android:layout_toRightOf = "@id/back"
    android:text = "haha"
    android:textColor = "#6600ff"
    android:textSize = "18sp" />
<TextView
    android:id = "@+id/systemtime"
    android:layout_width = "wrap_content"
    android:layout_height = "wrap_content"
    android:layout_centerVertical = "true"
    android:layout_marginRight = "20dp"
    android:layout_toLeftOf = "@+id/batteryimage"
    android:text = "08:45"
    android:textColor = "#6600ff"
    android:textSize = "18sp" />
<ImageView
    android:id = "@+id/batteryimage"
    android:layout_width = "wrap_content"
    android:layout_height = "wrap_content"
    android:layout_alignParentRight = "true"
```

```xml
        android:layout_gravity = "right"
        android:layout_marginRight = "20dp"
        android:layout_marginTop = "10dp"
        android:adjustViewBounds = "false"
        android:paddingRight = "0dp"
        android:src = "@drawable/battery_level4" />
</RelativeLayout>
<RelativeLayout
    android:id = "@+id/media_controller"
    android:layout_width = "match_parent"
    android:layout_height = "wrap_content"
    android:layout_alignParentBottom = "true"
    android:gravity = "bottom" >
    <LinearLayout
        android:id = "@+id/linearLayout"
        android:layout_width = "match_parent"
        android:layout_height = "wrap_content"
        android:layout_alignParentTop = "true"
        android:gravity = "center"
        android:orientation = "horizontal"
        android:paddingTop = "4dip" >
        <TextView
            android:id = "@+id/time_current"
            android:layout_width = "wrap_content"
            android:layout_height = "wrap_content"
            android:layout_gravity = "center_horizontal"
            android:paddingEnd = "4dip"
            android:paddingStart = "4dip"
            android:paddingTop = "4dip"
            android:textColor = "#6600ff"
            android:textSize = "14sp"
            android:textStyle = "bold" />
        <SeekBar
            android:id = "@+id/mediacontroller_progress"
            style = "?android:attr/progressBarStyleHorizontal"
            android:layout_width = "0dip"
            android:layout_height = "32dip"
            android:layout_alignParentEnd = "true"
            android:layout_alignParentStart = "true"
            android:layout_weight = "1" />
        <TextView
            android:id = "@+id/time"
```

```xml
        android:layout_width = "wrap_content"
        android:layout_height = "wrap_content"
        android:layout_gravity = "center_horizontal"
        android:paddingEnd = "4dip"
        android:paddingStart = "4dip"
        android:paddingTop = "4dip"
        android:textColor = "#6600ff"
        android:textSize = "14sp"
        android:textStyle = "bold" />
    </LinearLayout>
    <LinearLayout
        android:layout_width = "match_parent"
        android:layout_height = "wrap_content"
        android:layout_below = "@id/linearLayout"
        android:gravity = "center"
        android:orientation = "horizontal" >
    <ImageButton
        android:id = "@ + id/prev"
        style = "@style/MediaButton.Previous" />
    <ImageButton
        android:id = "@ + id/rew"
        style = "@style/MediaButton.Rew" />
    <ImageButton
        android:id = "@ + id/pause"
        style = "@style/MediaButton.Play" />
    <ImageButton
        android:id = "@ + id/ffwd"
        style = "@style/MediaButton.Ffwd" />
    <ImageButton
        android:id = "@ + id/next"
        style = "@style/MediaButton.Next" />
    </LinearLayout>
    </RelativeLayout>
    </RelativeLayout>
```

上面代码中,播放控制按钮中有用到样式 style,我们将对应的代码放在项目的 values 文件夹中的 styles.xml 文件中,如代码 9-23 所示。

【代码 9-23】 styles.xml 代码

```xml
<resources>
    <style name = "AppBaseTheme" parent = "android:Theme.Light">
    </style>
    <style name = "AppTheme" parent = "AppBaseTheme">
```

```xml
    </style>
    <style name = "MediaButton" >
        <item name = "android:background" >@null</item>
        <item name = "android:layout_width" >71dip</item>
        <item name = "android:layout_height" >52dip</item>
    </style>
    <style name = "MediaButton.Previous" >
        <item name = "android:src" >@drawable/pre</item>
    </style>
    <style name = "MediaButton.Next" >
        <item name = "android:src" >@drawable/next</item>
    </style>
    <style name = "MediaButton.Play" >
        <item name = "android:src" >@drawable/play</item>
    </style>
    <style name = "MediaButton.Ffwd" >
        <item name = "android:src" >@drawable/ff</item>
    </style>
    <style name = "MediaButton.Rew" >
        <item name = "android:src" >@drawable/rew</item>
    </style>
    <style name = "MediaButton.Pause" >
        <item name = "android:src" >@drawable/pause</item>
    </style>
</resources>
```

准备工作已经就绪，接着开始改造 MyMediaController 这个类。首先将系统自带的 MediaController 这个类的代码全部拷贝到 MyMediaController 中，把类的名字改为 MyMediaController，并修改所有的构造方法名字。接着开始定义一些成员变量，如代码 9-24 所示。

【代码 9-24】 MyMediaController.Java 部分代码

```java
public class MyMediaController extends FrameLayout {
    …//这里省略与 MediaController 类一样的代码
    private TextView        mPlayingVideoNameTv;  //当前正在播放的视频文件名称
    private TextView        mSystemTime;  //当前系统时间
    private ImageView       mBattery;  //手机当前电量
    private ImageButton     mBackBtn;  //后退按钮
    private String          playingVideoName;
    private int[] batteryImages = new int[]{R.drawable.battery_level0,
        R.drawable.battery_level1, R.drawable.battery_level2, R.drawable.battery_level3,
        R.drawable.battery_level4};  //电池电量对应图片
```

…//这里省略与MediaController类一样的代码

由于修改了视频播放界面的布局文件,所以需要修改MediaController类相关的方法,如makeControllerView()和initControllerView(View v)方法,修改后的代码如代码9-25所示。

【代码9-25】makeControllerView()和initControllerView(View v)代码

```
protected View makeControllerView() {
    LayoutInflater inflate = (LayoutInflater) mContext.getSystemService(Context.LAYOUT_INFLATER_SERVICE);
    mRoot = inflate.inflate(R.layout.media, null); //将自定义的media.xml作为播放界面
    initControllerView(mRoot);

    return mRoot;
}
private void initControllerView(View v) {
    mPlayingVideoNameTv = (TextView) v.findViewById(R.id.playingvideoname);
    mPlayingVideoNameTv.setText(playingVideoName);
    mSystemTime = (TextView) v.findViewById(R.id.systemtime);
    mBattery = (ImageView) v.findViewById(R.id.batteryimage);
    mBackBtn = (ImageButton) v.findViewById(R.id.back);
    if(mBackBtn != null){
        mBackBtn.setOnClickListener(mBackListener);
    }
    mPauseButton = (ImageButton) v.findViewById(R.id.pause);
    if (mPauseButton != null) {
        mPauseButton.requestFocus();
        mPauseButton.setOnClickListener(mPauseListener);
    }
    mFfwdButton = (ImageButton) v.findViewById(R.id.ffwd);
    if (mFfwdButton != null) {
        mFfwdButton.setOnClickListener(mFfwdListener);
        if (!mFromXml) {
            mFfwdButton.setVisibility(mUseFastForward ? View.VISIBLE : View.GONE);
        }
    }
    mRewButton = (ImageButton) v.findViewById(R.id.rew);
    if (mRewButton != null) {
        mRewButton.setOnClickListener(mRewListener);
        if (!mFromXml) {
            mRewButton.setVisibility(mUseFastForward ? View.VISIBLE : View.GONE);
```

```
        }
    }
    mNextButton = (ImageButton) v.findViewById(R.id.next);
    if (mNextButton ! = null && ! mFromXml && ! mListenersSet) {
    mNextButton.setVisibility(View.GONE);
    }
    mPrevButton = (ImageButton) v.findViewById(R.id.prev);
    if (mPrevButton ! = null && ! mFromXml && ! mListenersSet) {
    mPrevButton.setVisibility(View.GONE);
    }
    mProgress = (ProgressBar) v.findViewById (R.id.mediacontroller_progress);
    if (mProgress ! = null) {
    if (mProgress instanceof SeekBar) {
    SeekBar seeker = (SeekBar) mProgress;
    seeker.setOnSeekBarChangeListener(mSeekListener);
    }
    mProgress.setMax(1000);
    }
    mEndTime = (TextView) v.findViewById(R.id.time);
    mCurrentTime = (TextView) v.findViewById(R.id.time_current);
    mFormatBuilder = new StringBuilder();
    mFormatter = new Formatter(mFormatBuilder, Locale.getDefault());
    installPrevNextListeners();
}
```

在 initControllerView（View v）方法中的第 7 行代码，我们给后退按钮 mBackBtn 设置了监听器对象 mBackListener，mBackListener 对象的创建过程如代码 9 – 26 所示。

【代码 9 – 26】创建 mBackListener 对象

```
private View.OnClickListener mBackListener = new View.OnClickListener() {
    public void onClick(View v) {
        Activity ac = (Activity)mContext;
        ac.finish();
    }
};
```

在改造 MyMediaController 中遇到的一个棘手问题是，在 initFloatingWindow（）方法中，系统自带的 MediaController 类是通过 PolicyManager. makeNewWindow（mContext）这句代码来获得 mWindow 对象的，但是 PolicyManager 这个类属于 Android 内部使用，我们用不了。那怎么办呢？我们可以通过 Java 反射技术来解决这个问题。我们新写一个静态方法 getPolicyWindow 来获取 mWindow 对象，如代码 9 – 27 所示。

【代码 9 – 27】getPolicyWindow（）方法

```
public static Window getPolicyWindow(Context context) {
```

```
        Window window = null;
        try {
            Class<?> policyManagerClass = Class
    .forName("com.android.internal.policy.PolicyManager");
            Class<?>[] parMakeNewWindow = { Context.class };
            Method makeNewWindow = policyManagerClass.getDeclaredMethod(
                "makeNewWindow", parMakeNewWindow);
            Object[] args = { context };
            window = (Window) makeNewWindow.invoke(null, args);
        } catch (Exception e) {
            e.printStackTrace();
        }
        return window;
}
```

有了这个方法后,我们就可以将 initFloatingWindow( )方法的第二行代码〔即 mWindow = PolicyManager.makeNewWindow(mContext)〕用 mWindow = getPolicyWindow(mContext)替换以编译通过并达到同样的效果。

由于自定义播放界面要显示系统当前时间(参考图 9-24),那怎么获取系统时间并实时更新呢?这里就要用到 Java 中的线程技术和 Android 中的 Handler 技术了。我们创建一个线程对象和一个 Handler 对象,线程对象每隔一秒给 Handler 对象发送一个消息(Message),Handler 对象收到这个消息后将获取系统时间并将其显示在相应的 TextView 控件上(即之前已经定义了的 mSystemTime),如代码 9-28 所示。

【代码 9-28】创建 Handler 对象及线程

```
private Handler timeHandler = new Handler() {
@Override
public void handleMessage(Message msg) {
    super.handleMessage(msg);
    switch (msg.what) {
    case 1:
        long sysTime = System.currentTimeMillis();//获取系统时间
        CharSequence sysTimeStr = DateFormat.format("hh:mm", sysTime);
        String time = sysTimeStr.toString();
        Log.i(TAG, "get12Or24Format() = " + get12Or24Format());
        if("24".equals(get12Or24Format())){
            String hh = (Integer.parseInt(sysTimeStr.subSequence(0,2).toString())
+ 12) + "";
            if("24".equals(hh)){
                hh = "00";
            }
            time = hh + sysTimeStr.subSequence(2, sysTimeStr.length()).toString();
        }
```

```
            mSystemTime.setText(time);
            break;
        default:
            break;
        }
    }
};
//内部类 TimeThread
private class TimeThread extends Thread{
    @Override
    public void run() {
        while(true){
            try {
                Thread.sleep(1000);
                timeHandler.sendMessage(timeHandler.obtainMessage(1));
            } catch (InterruptedException e) {
                //TODO Auto-generated catch block
                e.printStackTrace();
            }
        }
    }
}
```

接下来就是 TimeThread 这个线程什么时候启动呢？我们可以在 MyMediaController 的构造方法中启动，如代码 9-29 所示。

【代码 9-29】 启动 TimeThread 线程

```
public MyMediaController(Context context, boolean useFastForward) {
    super(context);
    mContext = context;
    mUseFastForward = useFastForward;
    initFloatingWindowLayout();
    initFloatingWindow();
    new TimeThread().start();//启动线程
}
```

为了方便实现电池电量图片的实时更新和显示当前播放的视频文件的名称这两个需求，需要在 MyMediaController 类中添加几个方法，如代码 9-30 所示。

【代码 9-30】 MyMediaController 类中新增方法

```
public String getPlayingVideoName(){
    return playingVideoName;
}
public void setPlayingVideoName(String name){
    this.playingVideoName = name;
}
```

```java
public void setCurBatteryPower(int curPower){
    mBattery.setImageResource(batteryImages[curPower]);
}
```

至此，MyMediaController 类的改造就算大功告成了。那增量9-3是否就已经完成了呢？别急，由于我们使用了自定义的 VideoView 和 MediaController，所以我们重新需要修改 VideoPlayActivity 类的代码。首先，VideoPlayActivity 类的显示布局 videoplayer.xml 中的 VideoView 控件需要修改为自定义的 MyVideoView 控件，如代码9-31所示。

【代码9-31】修改后的 videoplayer.xml

```xml
<?xml version="1.0" encoding="utf-8"?>
<LinearLayout xmlns:android="http://schemas.android.com/apk/res/android"
    android:layout_width="match_parent"
    android:layout_height="match_parent"
    android:baselineAligned="false"
    android:gravity="center"
    android:background="#000000" >
<nfsysu.cn.videoplayer.view.MyVideoView
    android:id="@+id/videoview"
    android:layout_width="match_parent"
    android:layout_height="match_parent"
    android:layout_gravity="center_horizontal"
    />
</LinearLayout>
```

接着修改代码，最主要的是修改 oncreate() 方法，如代码9-32所示。

【代码9-32】VideoPlayActivity 类的 oncreate() 方法

```java
public class VideoPlayActivity extends Activity {
    private static final String TAG = "VideoPlayActivity";
    private MyVideoView videoView;
    private VideoDBService vds;
    private int position;
    private MyMediaController mediaController;
    @Override
    protected void onCreate(Bundle savedInstanceState) {
        super.onCreate(savedInstanceState);
        requestWindowFeature(Window.FEATURE_NO_TITLE);
        getWindow().setFlags(WindowManager.LayoutParams.FLAG_FULLSCREEN,
                WindowManager.LayoutParams.FLAG_FULLSCREEN);
        setContentView(R.layout.videoplayer);
        vds = new VideoDBService(this);
        mediaController = new MyMediaController(this);
        videoView = (MyVideoView) this.findViewById(R.id.videoview);
        mediaController.setAnchorView(videoView);
```

```java
mediaController.setPrevNextListeners(new OnClickListener() {
    public void onClick(View v) {
        Cursor cursor = vds.getCursorScrollData(0, vds.getCount());
        if (position == vds.getCount() -1) {
            position = 0;
        } else {
            position++;
        }
        cursor.moveToPosition(position);
        String path = cursor.getString(cursor.getColumnIndex("path"));
        Log.i(TAG, "path: " + path);
        mediaController.setPlayingVideoName(path.substring(path
            .lastIndexOf("/") + 1));
        videoView.setVideoURI(Uri.parse(path));
        mediaController.setCurBatteryPower(curPower);
        videoView.start();
    }
}, new View.OnClickListener() {
    public void onClick(View v) {
        Cursor cursor = vds.getCursorScrollData(0, vds.getCount());
        if (position == 0) {
            position = vds.getCount() -1;
        } else {
            position--;
        }
        cursor.moveToPosition(position);
        String path = cursor.getString(cursor.getColumnIndex("path"));
        Log.i(TAG, "path: " + path);
        mediaController.setPlayingVideoName(path.substring(path
            .lastIndexOf("/") + 1));
        videoView.setVideoURI(Uri.parse(path));
        mediaController.setCurBatteryPower(curPower);
        videoView.start();
    }
});
videoView.setOnCompletionListener(new OnCompletionListener() {
    public void onCompletion(MediaPlayer mp) {
        Cursor cursor = vds.getCursorScrollData(0, vds.getCount());
        if (position == vds.getCount() -1) {
            position = 0;
```

```
                } else {
                    position++;
                }
                cursor.moveToPosition(position);
                    String path = cursor.getString(cursor.getColumnIndex("path"));
                Log.i(TAG, "path: " + path);
                mediaController.setPlayingVideoName(path.substring(path
                    .lastIndexOf("/") + 1));
                videoView.setVideoURI(Uri.parse(path));
                videoView.start();
                mediaController.setCurBatteryPower(curPower);
            }
        });
        videoView.setMediaController(mediaController);
        String path = this.getIntent().getStringExtra("path");
        position = this.getIntent().getIntExtra("position", 0);
        mediaController.setPlayingVideoName(path.substring(path
            .lastIndexOf("/") + 1));
        videoView.setVideoURI(Uri.parse(path));
        videoView.start();
}
```

接下来我们需要实现增量9-3的最后一个需求,那就是实时更新电池电量。我们这里将电池电量分为5个级别,电量小于25%为0级别,电量大于等于25%但小于50%为1级别,依次类推。每一个级别用一张图片来表示。那怎么获取手机当前电量呢? Android系统是通过广播机制(Broadcast)来实现的,当电池电量发生变化时,系统会发出一个广播,这个广播中就含有手机当前电量等信息,所以我们只需注册一个广播接收者(BroadcastReceiver)接收系统发出的广播就可以获得电池电量的信息了。在VideoPlayActivity类中我们增加一个内部类BatteryBroadcastReceiver,该类继承BroadcastReceiver,同时定义相关的成员变量,如代码9-33所示。

【代码9-33】内部类BatteryBroadcastReceiver

```
private BatteryBroadcastReceiver batteryReceiver = new BatteryBroadcastReceiver();
private int curPower = 0;
private class BatteryBroadcastReceiver extends BroadcastReceiver {
    @Override
    public void onReceive(Context context, Intent intent) {
        //TODO Auto-generated method stub
        if (intent.getAction().equals(Intent.ACTION_BATTERY_CHANGED)) {//电池变化的广播
            int level = intent.getIntExtra("level", 0);
```

```
            int scale = intent.getIntExtra("scale",100);
            curPower = (level * 100 /scale) /25;
            mediaController.setCurBatteryPower(curPower);//设置电池电量图片
        }
    }
}
```

广播接收者(BroadcastReceiver)需要在系统中注册才能接收相应的广播,注册方式有两种,一种是静态注册,也就是在项目清单文件 AndroidManifest.xml 中注册;另一种是动态注册,这里采用动态注册的方式。在 VideoPlayActivity 类中的 onResume()中注册广播接受者对象(batteryReceiver),在 onPause()中注销该广播接受者对象,因此我们需要重写 VideoPlayActivity 类的这两个方法,如代码 9-34 所示。

【代码9-34】重写 VideoPlayActivity 类 onResume()和 onPause()方法
```
@Override
protected void onResume() {
    super.onResume();
    IntentFilter filter = new IntentFilter(Intent.ACTION_BATTERY_CHANGED);
    registerReceiver(batteryReceiver, filter);
}

@Override
protected void onPause() {
    super.onPause();
    unregisterReceiver(batteryReceiver);
}
```

至此,增量9-3的所有需求就全部实现了。

9.4 总结与回顾

通过以上章节的项目增量开发过程,我们最终实现了基于 Android 平台的视频播放器的预定项目目标。本章实现的视频播放器,用到了 Android 中的很多组件和控件,也用到了 Java 语言中的高级技术(如线程技术和反射技术),因而具有较高的参考价值和应用价值,特别是对于即将从事 Android 应用开发的初学者。然而,本系统在代码和功能的实现上仍然存在以下的一些问题,由于篇幅有限,本章不作详细分析和实现,读者可根据自身情况对以上实现的视频播放器系统进行修改和完善,以期达到更好的实现效果和应用价值:

(1)目前该视频播放器的功能不是很多,用户体验不够好,可以在后续版本中增加新的功能,如在视频播放界面增加通过上下滑动屏幕来调节音量、左右滑动屏幕来调节播放进度的功能。这个功能需要用到手势识别技术,读者可以参考其他书籍或视频来实现该功能。

(2)目前该视频播放器只支持 MP4 和 3GP 格式的视频解码播放,对于其他格式如AVI、WMV 的视频文件支持得不是很好。原因是我们采用的是 Android 自带的视频编解码技术。如果要支持更多格式的视频编解码以及生成视频缩略图等功能，我们需要移植其他开源的编解码库，如 FFmpeg。FFmpeg 是一套可以用来记录、转换数字音频、视频，并能将其转化为流的开源计算机程序。它采用 LGPL 或 GPL 许可证，提供了录制、转换以及流化音视频的完整解决方案，包含了非常先进的音频/视频编解码库 libavcodec，很多优秀的音视频播放器都用到了 FFmpeg 的编解码技术。

（3）当 SD 卡中视频文件很多时，视频列表显示界面可能会出现显示问题。解决的办法是可以采用分批加载技术。另一个值得改进的地方是可以将 SD 卡中的音频文件也纳入进来，也就是既可以管理和播放视频文件，同时也可以管理和播放音频文件。

（4）界面布局和用到的图片稍显粗糙，可以请具有良好美工基础的同学或同事重新设计界面和图片，以求达到更好的用户体验。

（5）通过解决以上问题可以大幅提高该视频播放器的应用价值、参考价值以及软件质量。但是，以上问题的解决需要参考和学习很多其他知识，需要投入很多时间和精力。所以，在软件项目的开发过程中，要具有良好的学习能力和团队合作能力，做好项目需求、设计比较合理的软件流程和架构，最终设计出一款成功的软件产品。

9.5 软件使用说明

9.5.1 开发环境

Java 集成开发环境：Eclipse 3.4 或以上。
Android SDK 4.2。

9.5.2 软件使用方法

软件运行环境：Android 4.2 以上的模拟器或 Android 手机。

软件部署：将项目目录下 bin 文件夹下的后缀名为 apk 的文件拷贝到 Android 手机或模拟器中，然后安装该 apk 文件，单击安装后生成的图标即可运行该程序。

9.6 本章小结

本章设计并实现了一个基于 Android 平台的视频播放器，在章节安排上首先对该系统进行需求分析，确定系统的目标和增量开发计划；最后按照增量开发的方法，按照设计的增量分阶段实现每个增量要求的目标。

视频播放器共实现了三个增量：搜索视频并列表显示、视频播放及控制功能的实现、自定义视频播放界面。每个增量都使用了 Android 和 Java 中相应的的技术。前面两个增量的实现相对比较简单，而第三个增量的实现比较复杂，用到了 Java 语言中的线程技术和反射技术，对线程技术和反射技术理解得不是很透彻的读者可以参考相关的 Java 书籍。

参考文献

［1］苑俊英，陈海山. Java 程序设计及应用——增量式项目驱动一体化教程［M］. 北京：电子工业出版社，2013.

［2］［美］Y. Daniel Liang 著；李娜 译. Java 语言程序设计：进阶篇（原书第 8 版）［M］. 北京：机械工业出版社，2011.

［3］赵满来. 可视化 Java GUI 程序设计：基于 Eclipse VE 开发环境［M］. 北京：清华大学出版社，2010.

［4］梁胜彬. Java 程序设计实例教程［M］. 北京：清华大学出版社，2011.

［5］耿祥义，张跃平. Java 程序设计教学做一体化教程［M］. 北京：清华大学出版社，2012.

［6］埃克尔. Java 编程思想 教材教辅与参考书 正版图书［M］. 北京：机械工业出版社，2007.

［7］刘京华. Java Web 整合开发王者归来（JSP + Servlet + Struts + Hibernate + Spring）［M］. 北京：清华大学出版社，2010.

［8］明日科技. Java Web 从入门到精通［M］. 北京：清华大学出版社，2012.

［9］贾蓓，镇明敏，杜磊 等. Java Web 整合开发实战：基于 Struts 2 + Hibernate + Spring［M］. 北京：清华大学出版社，2013.

［10］孙卫琴. Tomcat 与 Java Web 开发技术详解（第 2 版）［M］. 北京：电子工业出版社，2009.

［11］欧二强等. Java 编程手记：从实践中学习 Java［M］. 北京：清华大学出版社，2013.

［12］周颖，余胜威 等. 程序员的数学思维修炼［M］. 北京：清华大学出版社，2015.

［13］周颖，余胜威 等. 我和数学有约［M］. 北京：清华大学出版社，2015.

［14］［瑞典］Erik Hellman 著；丁志虎，武海峰. Android 编程实战［M］. 北京：人民邮电出版社，2014.

［15］黄永丽，王晓，孔美云. Android 应用开发完全学习手册［M］. 北京：清华大学出版社，2015.

［16］郭金尚等. Android 经典项目案例开发实战宝典［M］. 北京：清华大学出版社，2015.